POWDER METALLURGY EQUIPMENT MANUAL

3rd Edition

Editor
Samuel Bradbury

Prepared by the
Powder Metallurgy Equipment Association

Division of
Metal Powder Industries Federation
105 College Rd., E., Princeton, New Jersey 08540

First Published in 1968 and 1977, this third edition is completely revised to reflect current practices and the latest process equipment.

Library of Congress Catalog Card No.: 76-523-33

ISBN 0-918404-68-1

Printed in the United States of America

JL-2½-3/86

FOREWORD

Powder metallurgy (P/M) is an exacting technology which requires highly engineered equipment to meet the rigid specifications of the modern powder metallurgist and his customer.

P/M is a capital intensive technology offering economies in materials, labor and energy but exacting a demand on the equipment manufacturer to remain alert to changing needs and advancing industrial sophistication. The manual is published for this purpose.

The manual was prepared by members of the Powder Metallurgy Equipment Association, a division of the Metal Powder Industries Federation. The Federation is an international trade association serving the metal powder producing and consuming industries. Compiled by experts in the field, the manual describes the operation and principal characteristics of the different types of equipment used to mix metal powders, press them into shapes and to sinter those shapes. It does not attempt to detail the production of P/M products but serves primarily as a detail reference source for individuals already using and familiar with P/M fabricating equipment and the technology of powder metallurgy.

For information about specific pieces of equipment contact members of the Powder Metallurgy Equipment Association (PMEA). A directory containing manufacturers' catalogs and names and addresses of leading P/M equipment suppliers is available from PMEA on request. For information about the powder metallurgy process, suppliers of metal powders, P/M parts and other related products, industry standards and publications, contact the Metal Powder Industries Federation. Write to PMEA or the Federation at 105 College Road East, Princeton, New Jersey 08540. Telephone (609) 452-7700.

CONTENTS

INTRODUCTION TO THE P/M PROCESS

Powder metallurgy is a metal forming process for producing a variety of structural parts and bearings. Powders are mixed and then pressed in a die to the required shape. Next, the ejected compact is heated in a controlled atmosphere to bond the contacting surfaces of the metal particles and to obtain the desired properties of the part. This Manual covers mixing, pressing, tooling, sintering, and a section discussing additional operations including infiltration, joining, machining and heat treating that might be desirable. (fig. 1).

Most materials used are metals in powder form, which range in particle size from –80 to –325 mesh (–175 to –43 μm). Most metals can be pressed to useful shapes by this method. Aluminum, iron, alloy steel, stainless steel, copper, brass, bronze, nickel, and nickel alloys are widely used. However, carbon, graphite, plastics, carbides, ceramics, as well as mixtures of metallic and non-metallic powders, are shaped commercially also by P/M techniques.

Early powder metallurgy parts were bearings and washers—simple shapes with relatively low mechanical properties. Improvements in compacting systems, tooling, powder alloys and sintering furnaces have provided the means to produce larger, more intricate and stronger parts. Structural shapes with flanges, hubs, cores, counterbores, and combinations of these, are commonplace.

P/M components offer many inherent advantages which cannot be attained by any other metalworking processes. For example:

Precise Control: The powder metallurgist is able to control his product from the pure powder to the finished part. He can create the material and produce a finished product. Thus, the properties and characteristics of the end product can be suited to the demands of the application. Eliminated are poor finishes, uneven internal stresses, impurities, inclusions, unworkable tolerances, or the many other factors which might affect the rate of part production or the quality of the finished product. The manufacturer who uses powder metallurgy parts is assured of uniformity and optimum performance characterists.

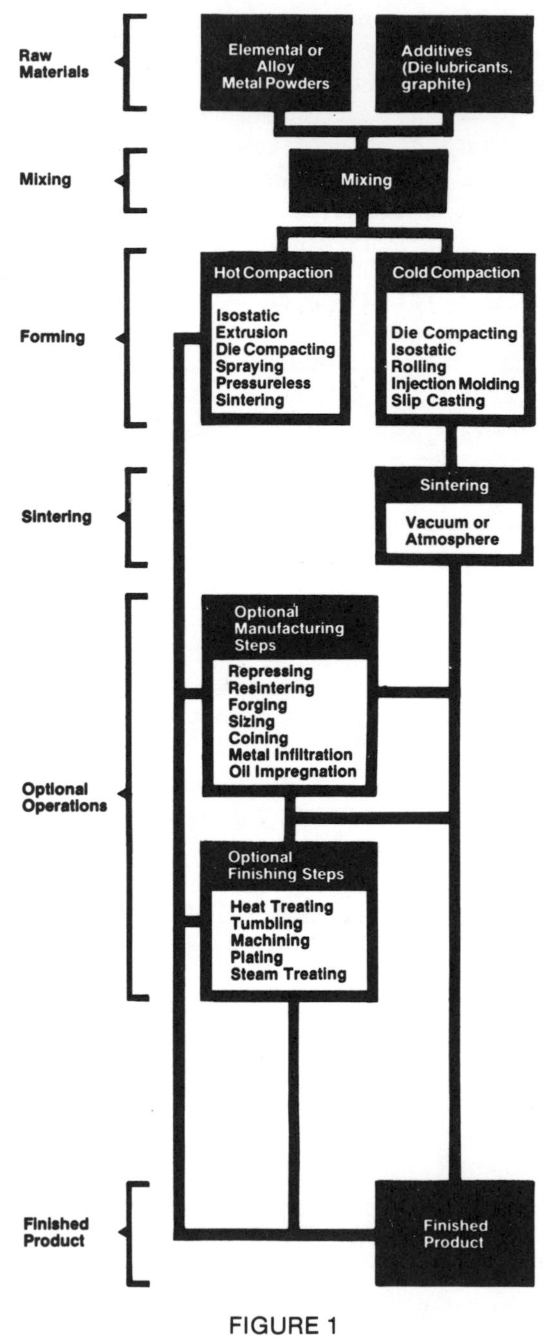

FIGURE 1
The powder metallurgy process.

"Custom-Made" Compositions: Immiscible materials can be combined to produce specific properties not attainable by pyrometallurgy. Dissimilar metals, non-metallics, plus materials of widely different characteristics can be compacted into parts which have unique properties. Ceramics can be blended with metals. Carbon and copper are compacted to form electrical brushes with wear resistance plus electrical conductivity. Tungsten and silver are immiscible by fusion metallurgy, but their characteristics are combined by powder metallurgy in the manufacture of electrical contacts and switch gear. Copper, tin, iron, lead and graphite are compacted into heavy-duty friction materials. The combining of materials through the use of powder techniques is limited only by research and experimentation.

Unusual Physical Properties: Physical properties can be varied from low density, highly porous filters, to high density parts with minimal porosity. Tensile strengths can be varied from low to very high. Dissimilar materials can be compacted in layers to attain a single part with individual properties on opposite surfaces.

Self-lubrication is another example of a unique property attainable only through the use of powder technology. The porosity in a part is controlled so that it can be impregnated with oil or other lubricants to become self-lubricating. Gears and other parts manufactured from metal powder possess sound damping characteristics which permit quieter operation. A powder metallurgy gear does not "ring," yet it can possess all the other features of a wrought metal gear.

Reduced Machining and Finishing: The experienced custom parts manufacturer can produce parts in a wide range of finished shapes and sizes which require no further handling and processing. Many gears, cams and intricately shaped parts which require expensive machining to produce from wrought stock can be made from metal powders. Counterbores, flanges, hubs, and holes can be formed when parts are compacted. Keyways, keys, D-shaped bores and other fastening devices can be made an integral part of the component. It is also common to combine two or more parts into a single unit when product design permits.

No Material Loss and Less Material Usage are Economic Benefits: With the elimination of further processing such as machining and finishing, parts are produced without scrap. The use of powder metallurgy parts reduces the need for inventories of bars, rods, plates, sheet and strip.

Reproducibility...The First is the Same as the Last: The dies in which powders are compacted are among the most repetitively accurate devices found in any mass production process. Each part made in a die duplicates the preceding part. Part deviation can take place only as the die wears, and with the use of carbide dies this becomes detectable only after tens of thousands of parts are produced.

An Energy Efficient Process: A detailed analysis of energy consumption showed that the total energy required to produce a P/M part is often 25% to 50% of that required to produce the same part by machining it from a wrought starting material.

PART

ONE

MIXING

10.00 MIXING

According to MPIF Standard 09* "mixing" of powders refers to "the thorough intermingling of powders of two or more materials." The term "blending" refers to "the thorough intermingling of powders of the same material." For the production of P/M parts we are concerned primarily with "mixing" a powdered die wall lubricant with a metal powder, graphite powder or with powders of several other metals. In the latter instance the graphite and metal powders diffuse with one another during sintering to become an alloy.
emptying of the mixer for each batch. For P/M operations the mixer stands are often made high enough to provide clearance beneath the mixer so it may be discharged into a moveable bin or hopper that will be placed directly over the compacting press feed.

Mixers are available from a laboratory size 0.5 ft^3 (0.014 m^3), 9″ (230mm) diameter to 2,000 $ft.^3$ (57 m^3) and 8 ft. (2.4 m) diameter.

11.10 TYPES OF MIXERS

11.11 Double Cone

Double cone mixers may be described as vertical cylinders with conical ends. In operation they rotate about a horizontal axis. This rotation imparts a continuous rolling motion which spreads and folds the powders as they move in and out of the conical areas. This action thoroughly mixes the powders with little or no change in the size or shape of the individual particles (figs. 10-1, 10-2).

One mixer manufacturer offers a double cone mixer on which one conical end is clamped to the cylindrical center. This end is fitted with wheels for easy transport from a loading station to the mixer and then to storage or to a compacting press. With several conical ends on hand, little time is lost in charging and discharging the mixer (fig. 10-3).

*Also ASTM B243.

FIGURE 10-1
Double cone mixer with special high stands for additional working clearance.

A third mixer design combines the double cone and long leg V (cross-flow) advantages into what is called the Slant Cone. Essentially the slant cone is a double cone mixer with an off-set or slanted cylindrical center band section to promote axial material flow. The units are designed with D/3 centerband height (1/3 of center band diameter as oposed to the D/4 or 1/4 diameter design) and sized for 55% fill levels or 45% void. Double cone and V-shape mixers produce average internal batch variations of one or two percent. The slant cone produces average variations of one-half to one percent in one half to one-third blending time (fig. 10-4).

11.12 V Mixer

V shaped mixers are constructed by joining two cylinders of equal length into a "V." As the "V" rotates about its horizontal axis the powder charge splits and refolds. This action is gentle but thorough enough to mix the ingredients with minimal change in shape or density of the powder particles (fig. 10-5).

A variation of the V mixer is available in which two cylinders of unequal

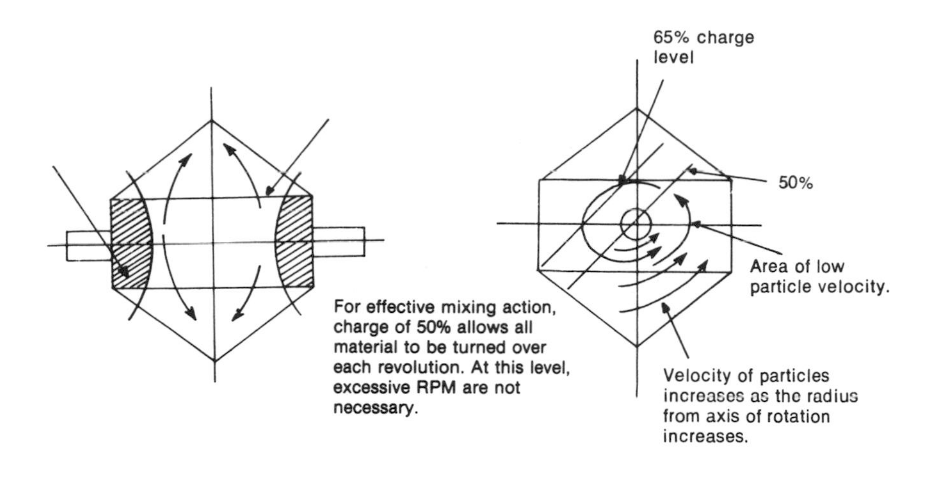

FIGURE 10-2
Double cone mixer-flow pattern.

FIGURE 10-3
Double cone mixer with portable hopper.

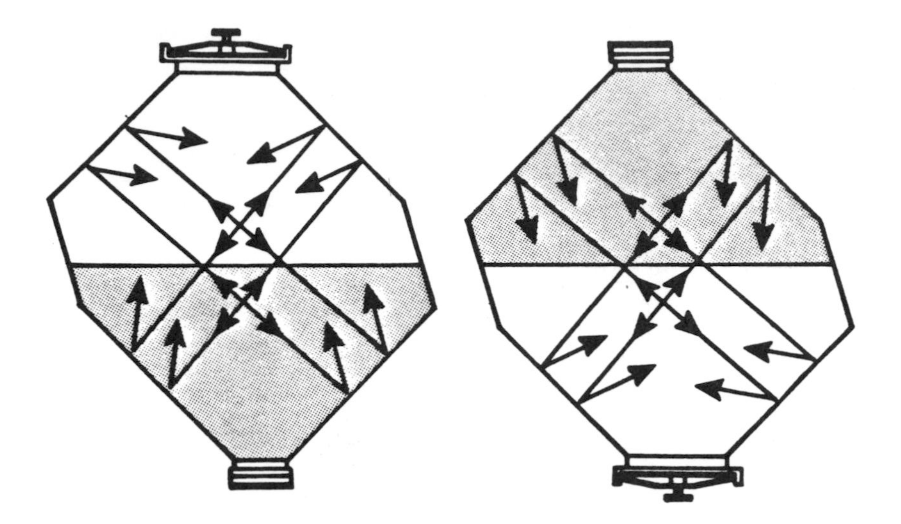

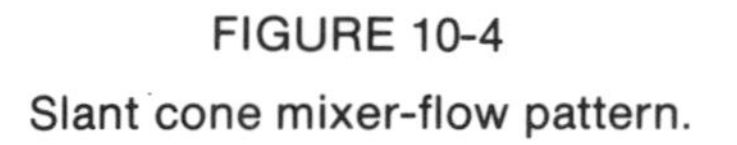

FIGURE 10-4

Slant cone mixer-flow pattern.

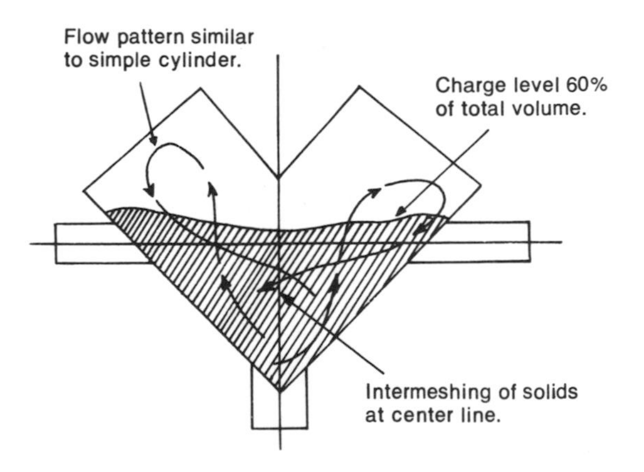

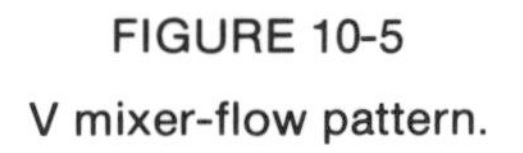

FIGURE 10-5

V mixer-flow pattern.

length are joined to form a "V." In such a mixer there is a strong axial flow of material in addition to the splitting and refolding action of the basic twin shell mixer. This axial flow helps to reduce mixing time (fig. 10-6).

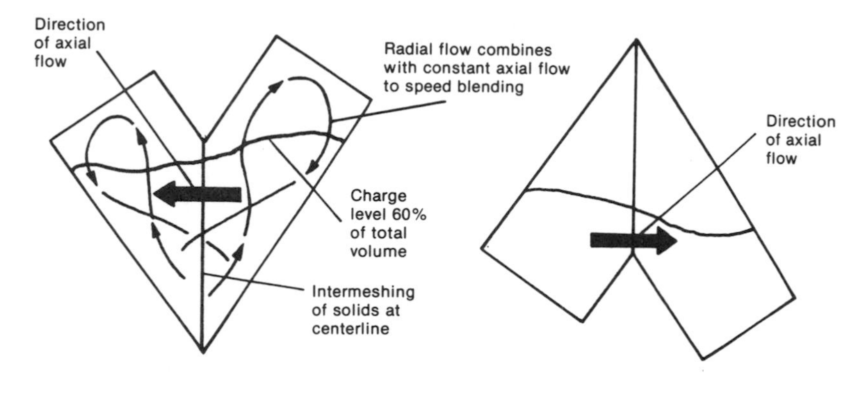

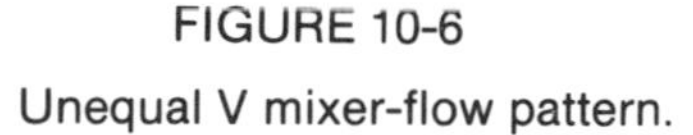
FIGURE 10-6
Unequal V mixer-flow pattern.

12:00 MIXING VARIABLES

12.11 Size of Mix

Mixing efficiency is influenced a great deal by the relationship of the volume of powder to the volume of the mixer. Efficiency is best when the powder volume is about 50% to 60% of the mixer volume. Mixer design, interior finish and mixing speed each has its effect on mixing efficiency. The optimum mixer load is usually determined by testing different volumes in a given mixer.

12.12 Mixing Time

Optimum mixing time may be from about 5 to 30 minutes but this can be determined only by experience with a given mixture in a particular mixer. The ideal is to mix the powders only as long as necessary to achieve a thorough mix and to maintain a uniform apparent density of the mix from batch to batch. Apparent density of the mix tends to increase with mixing time.

12.13 Powder Apparent Density

The term "apparent density" (A.D.) defines the weight/volume of a powder mass expressed in g/cm^3. It is important to control the A.D. of a powder during mixing because most P/M dies are designed to receive a specific weight/volume of powder to make a part to a specific pressed thickness at a specific pressed density. If the apparent density of the powder varies then obviously the thickness and/or the pressed density of the part may vary beyond allowable tolerances.

A.D. can be controlled by the length of mixing time and the mixing speed. A.D. usually increases with mixing time. A.D. can be decreased by adding a low A.D. powder to the mix before mixing is complete. MPIF Standard #4, "Determination of Apparent Density of Free-Flowing Metal Powders with Hall Apparatus" is a helpful reference and it details the procedure for determining the A.D. of a powder.

12.14 Powder Flow

Powder flow rate is a measure of how well a powder mix will flow into a die from an automatic press feeding device. Powder metallurgists depend on a uniform flow rate to fill the die completely with a uniform volume of powder from an automatic powder feed each cycle of the compacting press. Flow rate is expressed in seconds as the time required for a given volume of powder to flow through a specific orifice. MPIF Standard #3 "Determination of Flow Rate of Metal Powders Using the Hall Apparatus" is a useful reference.

PART TWO

COMPACTING

20.00 PRESSING

21.00 COLD PRESSING

The words pressing, compacting and briquetting when related to P/M are synonomous and imply the cold pressing of a measured amount of powder, in a die with one or more punches, to a densified form, called a green compact.

Compacting presses use mechanical or hydraulic means, or a combination of them, for the automatic and continuous cycle of powder feeding—compacting—ejecting the part from the die. The production rate can be from 200 to over 60,000 parts per hour, depending on the part sizes, their complexity and the type of press used.

21.11 Compacting and Ejecting Requirements

Pressing, stroke and depth of fill capacity are important considerations in selecting a pressing system. All press manufacturers list rated compacting capacities and stroking dimensions in their specification data sheets. The compacting tonnage listing describes the maximum force the press is capable of exerting during compression while maintaining acceptable life of the press components. Stroke is usually given in terms of punch travel while depth of die fill and ejection stroke are stated in inches, (mm).

There is little or no hydrostatic flow of metal powders during cold compacting. Therefore, it is desirable to support each level of the part with a separate punch or die member to maintain reasonably uniform pressed density throughout the part.

21.12 Pressing Force Required

The total load requirement of a press in tons is a function of the molding pressure of the material multiplied by the projected area of the part to be compacted. The molding pressure requirements are determined by the desired final density of the part (see Table 21-I). This type of data is normally furnished by all powder producers for their various products.

TABLE 21-I
Pressing Force Requirements and Compression Ratios for Various Materials

Type of Material	Pressing Force tsi	Pressing Force N/mm^2	Compression Ratio
Aluminum	5 to 20	70-280	1.5 to 1.9:1
Brass	30 to 50	415-690	2.4 to 2.6:1
Bronze	15 to 20	205-230	2.5 to 2.7:1
Copper-Graphite brushes	25 to 30	345-514	2.0 to 3.0:1
Carbides	10 to 30	140-415	2.0 to 3.0:1
Ferrites	8 to 12	110-165	3.0:1
Iron Bearings	15 to 25	205-345	2.2:1
Iron Parts: low density	25 to 30	345-415	2.0 to 2.4:1
medium density	30 to 40	415-550	2.1 to 2.5:1
high density	35 to 60	430-825	2.4 to 2.8:1
Iron Powder cores	10 to 50	140-690	1.5 to 3.5:1
Tungsten	5 to 10	70-140	2.5:1
Tantalum	5 to 10	70-140	2.5:1

The above pressing force requirements and compression ratios are approximations and will vary with changes in chemical, metallurgical, and sieve characteristics of the powder; with the amount of binder or die lubricants used; and with mixing procedures.

If for example a part having a projected area of 1.25 in^2 (807 mm^2) is to be pressed using a matrial requiring 30 tsi (414 N/mm^2) to provide the required density, a simple formula shows the compacting force required.

Unit compacting force x projected area = total press force.

$$30 \text{ tsi} \times 1.25 \text{ in}^2 = 37.5 \text{ tons}$$

or: *(Eq. 21-1)*

$$(414 \text{ N/mm}^2) \times (807 \text{ mm}^2) = 34\text{t}$$

21.13 Ejection Force Required

Ejection capacity is an important rating of a press. It is usually stated as tons by the manufacturer of the press. Some press manufacturers, because of the design of their machines, divide the rated ejection load. They will list an initial ejection capacity equivalent to compressive force for a pre-defined ejection breakaway stroke, which is usually between 1/32 in (0.79 mm) and 1/2 in (12.7 mm) at the start of the stroke. A reduced sustained ejection force, normally between 25% to 50% of the maximum compressive force, will also be listed. The ejection force required depends on punch, core rod and die side wall contact areas, tooling material, tool surface finish, and the amount and type of lubricants. For a rough calculation of ejection force required, multiply the part contact area with the die and core rod surfaces by the ejection breakaway pressure of the material.

21.14 Fill Stroke Required

During compaction in the die, the volume of the filled powder is reduced. The ratio of the volume of loose powder (apparent density) to the volume of the compacted part is the compression ratio of the powder (fig. 21-1 and table 21-I). The most common way of expressing density is g/cm³.

FIGURE 21-1

In practice, if a part of 2 in (51 mm) thick is to be pressed to a green density of 6.25 g/cm³, and if the powder selected has a compression ratio of 2.5:1, then a 5 in (127 mm) die fill is needed for the part. Figure 21-2 illustrates a straight wall die, lower punch, and upper punch, with loose powder in the die cavity in the fill position. The distance from the top surface of the die to the top surface of the lower punch is defined as the depth of fill. The amount of powder fill required to produce a part is determined by multiplying the finished part thickness by the compression ratio of material to be compacted to the required green density.

21.15 Ejection Stroke Required

Ejection stroke is measured as the distance from the top of the lower punch to the top of the die in the compression or pressed position (fig. 21-2, compression). The press at this stage must have enough stroking capability either to push the part up to the top surface of the die or to strip the die down to a point where the top surface of the die is flush with the top of the lower punch. At this point of the cycle the part is removed by the advancing feed shoe. The normal ejection stroke is equivalent to the thickness of the part plus the total amount of upper punch penetration into the die.

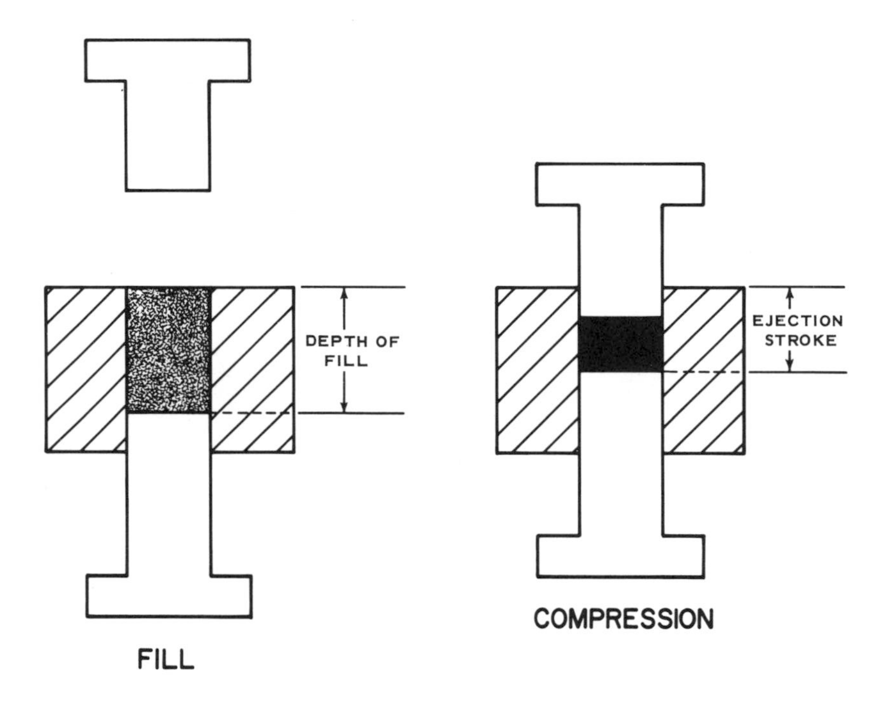

FIGURE 21-2

21.16 Part Classifications

P/M parts usually are classified by evaluating the complexity of part design on a range of I through IV, as shown by figs. 21-3 to 21-6. The more complex parts in fig. 21-6 require more complex press and tooling systems to achieve uniform density throughout all levels of the part.

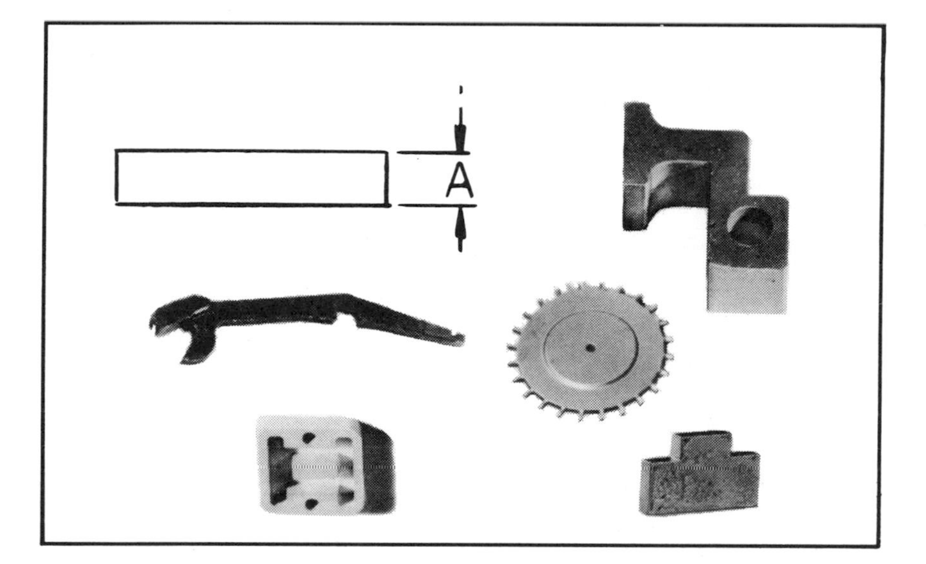

FIGURE 21-3
Class I parts.

Class I parts are thin, one-level parts of any contour that can be pressed with a force from one direction. The A dimension is generally limited to a maximum of 0.250 in (6.35 mm).

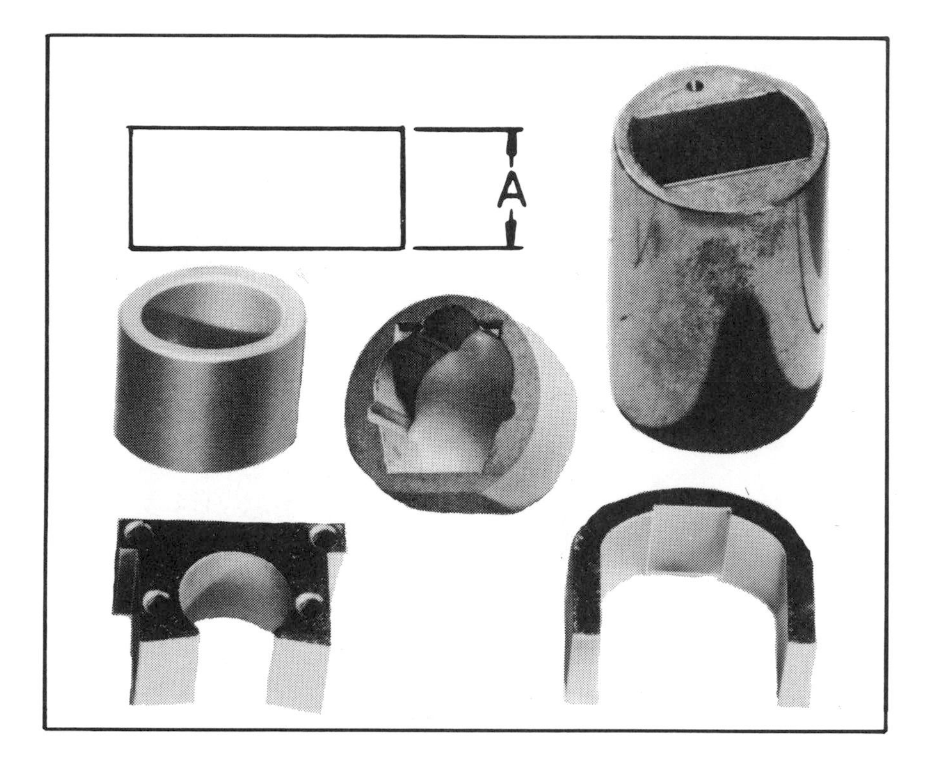

Class II parts.

FIGURE 21-4
Class II parts are one-level parts of any thickness and contour that must be pressed with forces from two directions.

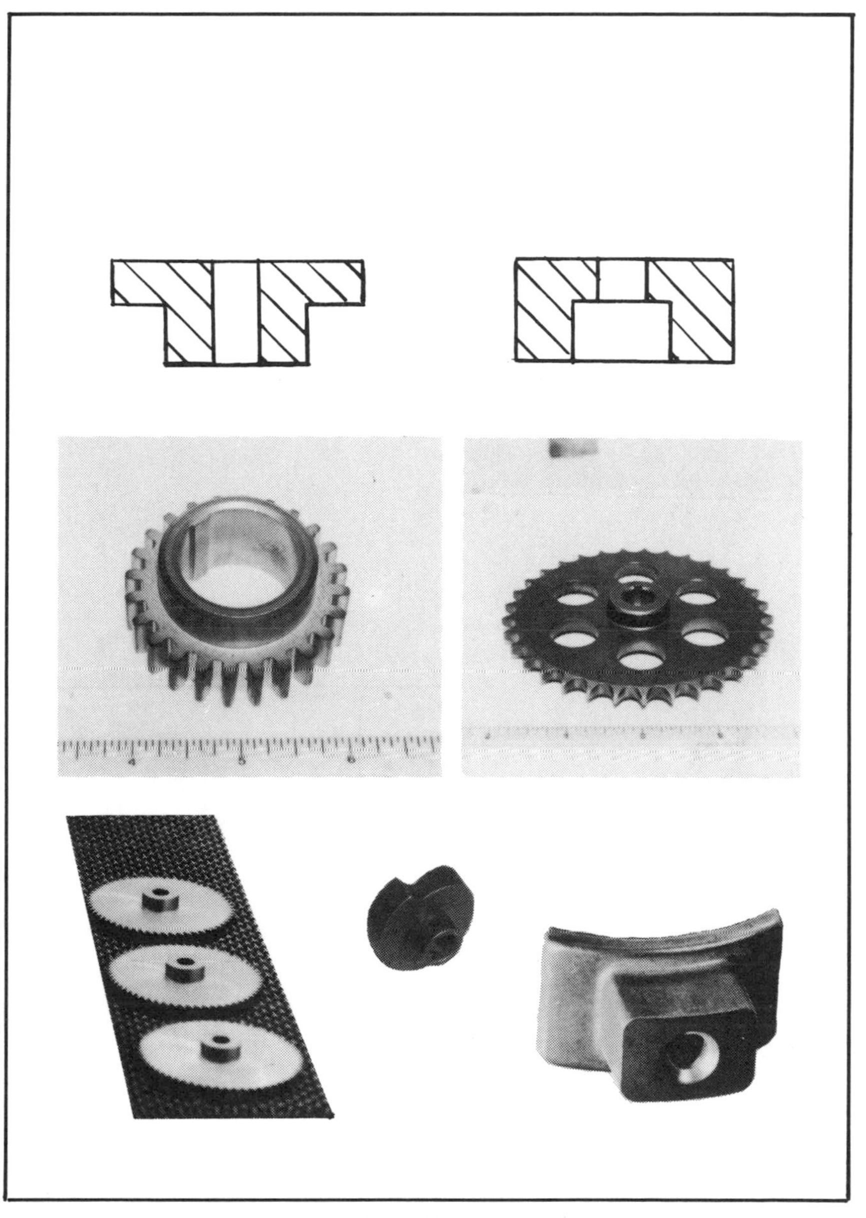

Class III parts.

FIGURE 21-5
Class III parts are two-level parts of any thickness and contour that must be pressed with forces from two directions.

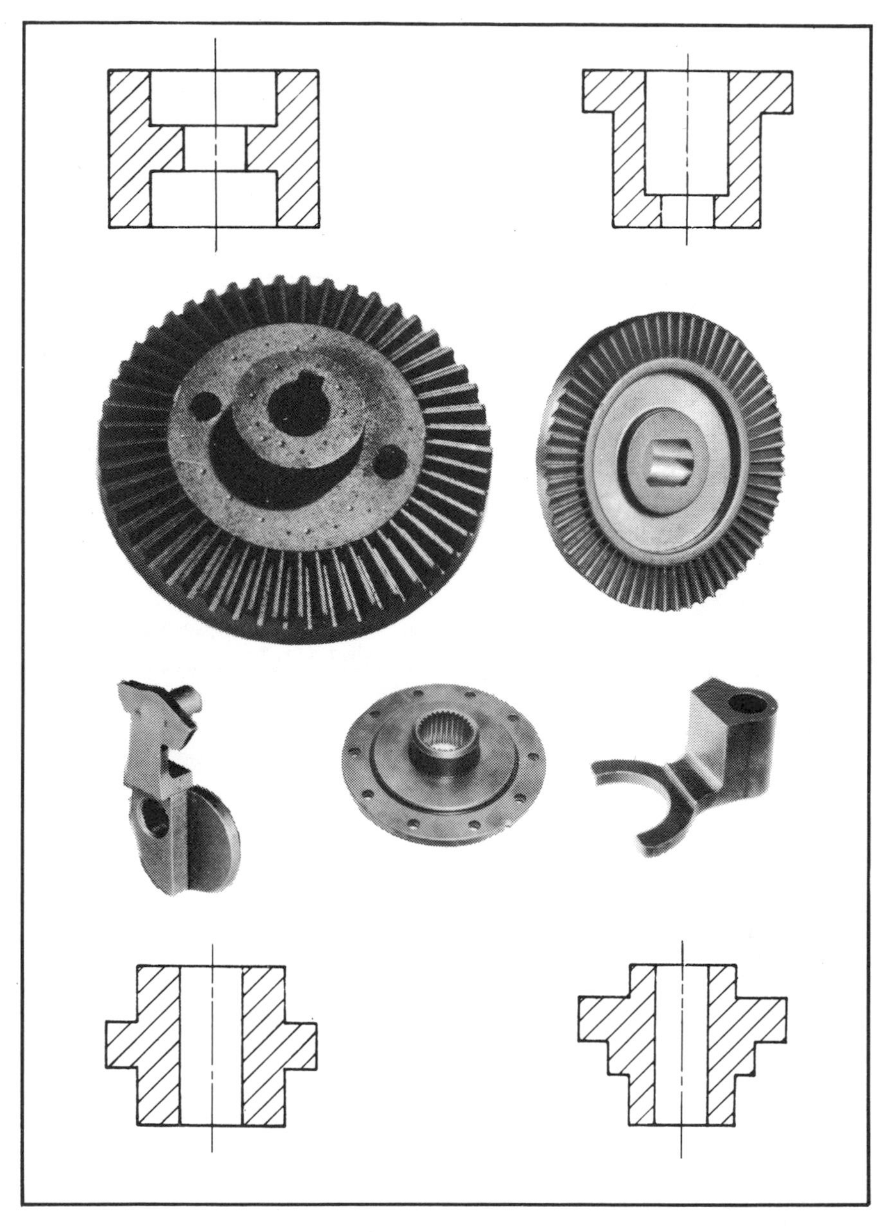

Class IV parts.

FIGURE 21-6
Class IV parts are multi-level parts of any thickness and contour that must be pressed with multiple forces from two directions.

21.17 Neutral Axes

Compaction begins as the upper punch enters the die to compress the powder. The density of the material normally increases first at the face of the upper punch. Friction develops between the partially compacted powder and sidewalls of the die, between the individual particles of the powder repacking to reduce voids in the early stages of compacting, between particle to particle from plastic flow as the particles are distored in the later stage of compaction, and between the punch faces and powder particles due to sideways movement of the grains.

Since powders do not flow hydrostatically under pressure, the friction generated absorbs part of the force being applied by the upper punch. This loss of force is not important when pressing thin parts. When thicker parts are pressed, force must be applied to both the top and bottom of the part to compensate for some of this frictional loss. The ability to transmit force throughout the powder being compacted determines the uniformity of density of the green compact. Uniform density is important to ensure dimensional control during sintering. Generally, uniform density is not reached when the part length/diameter ratio exceeds 3:1.

Under pressure, powders will not flow from one part level to another. Therefore, when parts of more than one level are being pressed, a separate pressing force must be supplied for each level. The thickness of each level will determine whether the pressing force must be applied from one direction or from both directions. On parts with more than one level, the pressing forces must be applied to each level simultaneously to assure density uniformity.

The neutral axis is the low density plane perpendicular to the direction of pressing. The plane is present in the part because the powder tends to resist flowing under pressure, which results in a lesser densification of the material farthest away from the forming element of the tooling. Placement of this plane in the part is controlled by the relative tooling motions. Normal positional location is usually set at the center of the part. When there is more than one level being compacted, there will be a neutral axis present for each part level (fig. 21-7).

21.20 Pressing Motions

21.21 Single Motion Pressing

Single action compacting includes a die to form the outer contour of the part, an upper punch to form the top surface of the part, and a lower punch to form the bottom surface of the part. It can also include core rods to form any through holes. This system is usually limited to compacting relatively thin Class I parts.

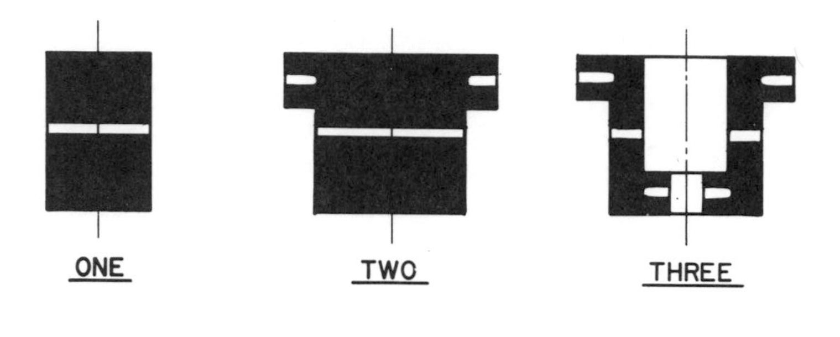

FIGURE 21-7

Neutral axes.

A. Upper punch pressing: During the compacting portion of the pressing cycle, the die, lower punch, and core rod if any, remain stationary. In operation the upper punch enters the die and compresses the powder against the stationary lower punch and inner surface of the die (fig. 21-8). Because the force applied by the press is from one direction only, it is called a single motion press. Ejection of the compacted part can be accomplished in either of two ways: the die and core rod remain stationary while the lower punch raises the part from the die cavity, or the lower punch remains stationary while the die and core rod are lowered to a point where the die table surface is flush with the top of the stationary punch.

B. Anvil pressing: A specialized form of single motion press is the anvil type. It includes a die to form the part contour, and a lower punch to form the bottom surface of the part. An upper punch is not used, since the top surface of the part is formed by an anvil or flat surface which slides over the filled die to level the powder. The anvil is then clamped over the die opening during compression.

Compaction can be by either of two methods. One uses movement of the lower punch for fill, compression, and ejection of the part, while the die table remains stationary (fig. 21-9). The other uses the lower punch as a fixed reference, while the die table is moved into positions for fill, compression, and ejection during the press cycle (fig. 21-10). Both actions can be defined as single pressing, since the compaction force is applied from only one direction. This system does offer the advantage of one less tooling member (upper punch), thereby minimizing the tool set costs and set-up time required for upper punches. Multiple cavity tooling with this system can provide additional machine production capacity.

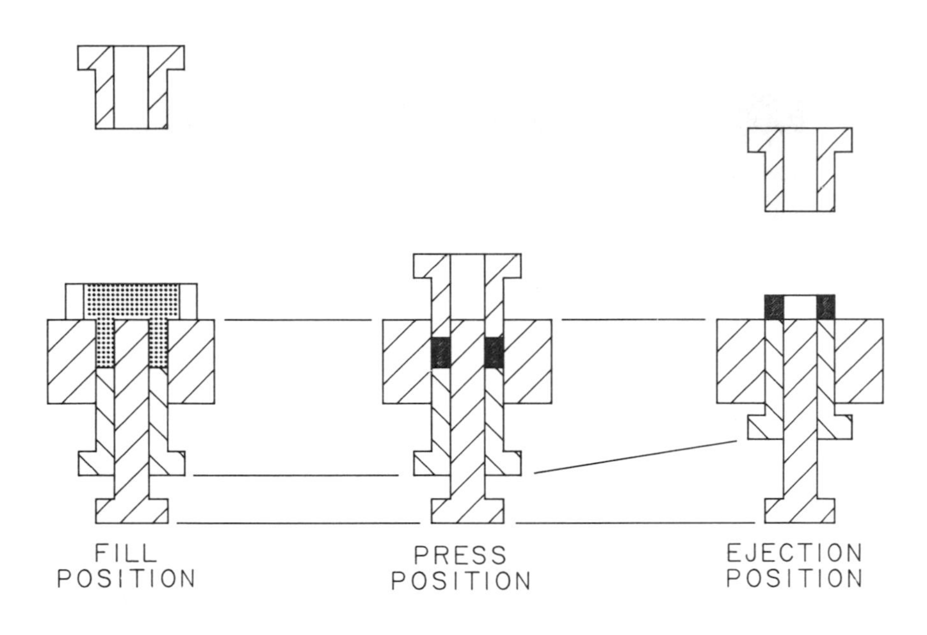

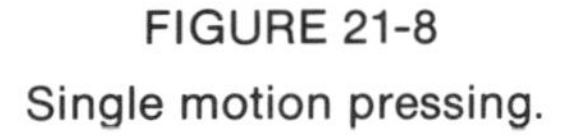

FIGURE 21-8
Single motion pressing.

21.22 Double Motion Pressing

This system is used primarily for the production of Class I and II parts, and certain Class III parts. It will produce parts with a more uniform density than single motion pressing.

A. Upper and lower punch pressing: The double motion opposed moving punch system applies force to the powder simultaneously from top and bottom through movement of the upper and lower punches (fig. 21-11). The lower punch normally has separately adjustable positions for powder fill, compression, and ejection. The upper punch is also independently adjustable to control the depth of penetration into the die. These adjustments permit positive control to position the neutral axis properly. Stationary and movable core rods can be used when parts with through or blind holes are being produced.

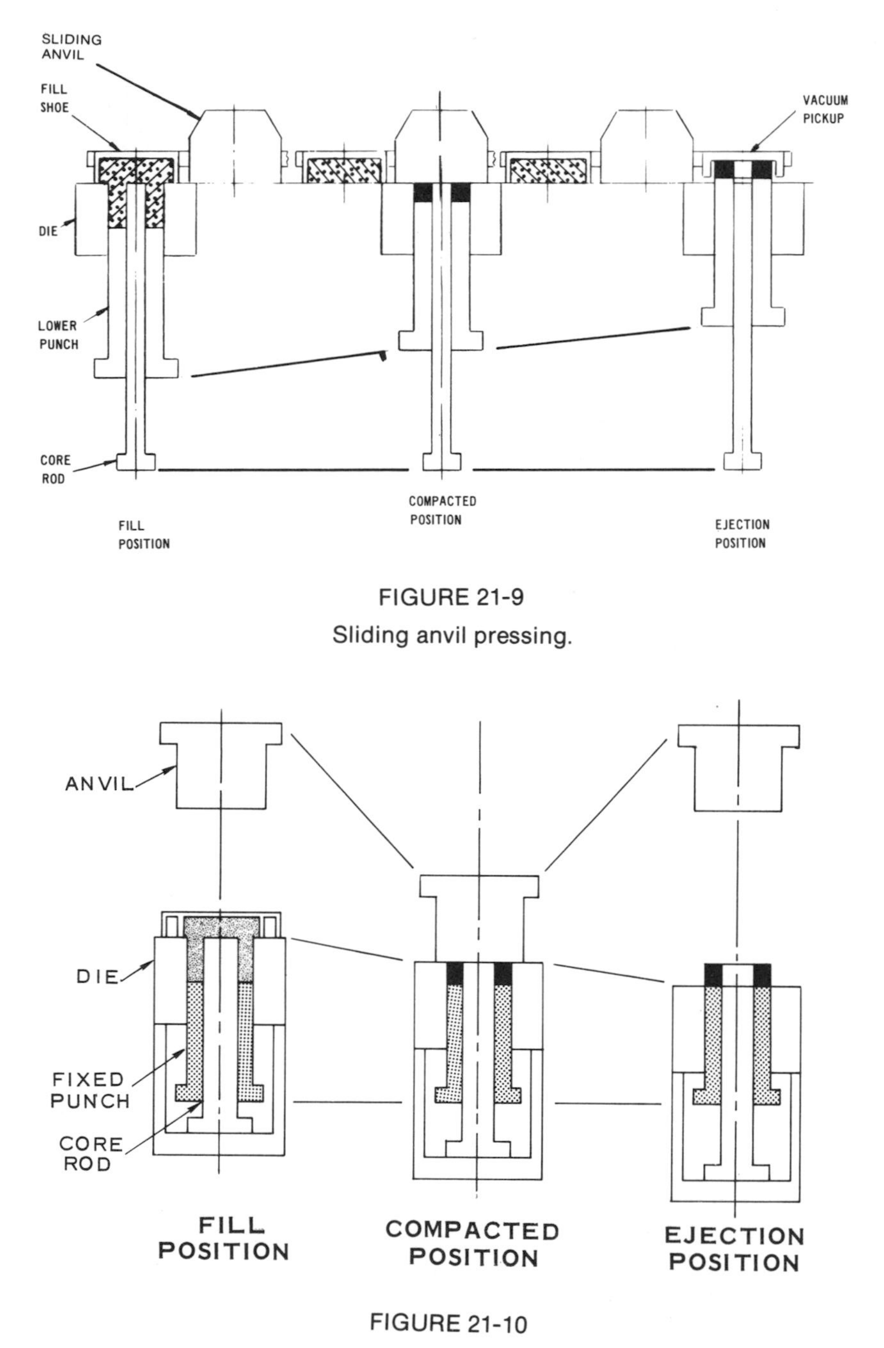

FIGURE 21-9

Sliding anvil pressing.

FIGURE 21-10

Anvil withdrawal pressing.

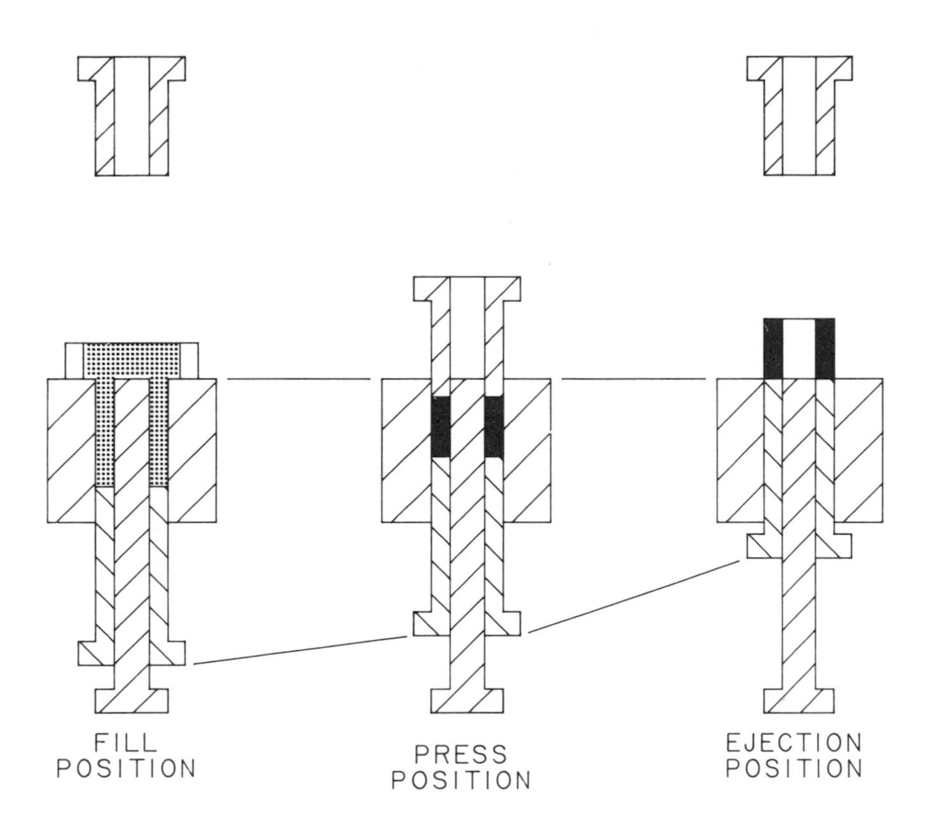

FIGURE 21-11

Double motion pressing.

B. Upper punch pressing with floating die: This system normally consists of a moving upper punch, moving die table or die element, and a stationary lower punch which is adjustable for fill. The die table or die element is held in position during the fill portion of the pressing cycle by a spring, air cylinder, or hydraulic cylinder. First, the upper punch moves into the die cavity and friction is developed between the powder and the inside surface of the die. When the frictional forces overcome the die supporting or counterbalancing force, the die descends at the same speed as the upper punch. This die movement has the same effect as a movable lower punch, thus applying pressure on the powder from both top and bottom (fig. 21-12). Normally a part pressed by this system will have density variations determined by the supporting force of the die. The neutral axis might not be located at the center of the part.

The density variation can be overcome if an outside force, greater than the die support force, is applied to the top surface of the die table as the upper punch enters the die cavity. The equalizing force is applied so that the die will move downward at a rate equal to 1/2 the speed of the descending upper punch. The resulting uniform pressure from both directions produces a part with the neutral axis in the center.

After compaction, the die remains in contact with the part, while the lower punch moves up to push the part out of the die cavity for ejection.

The core rod in its simplest arrangement is stationary. The core rod can have a separate fill and compacting position, completely independent of other press or tooling motions. The core rod can also be arranged so that it is free to move upward during ejection. The compacted part expands upon ejection from the die, freeing the core rod, thus reducing core rod wear.

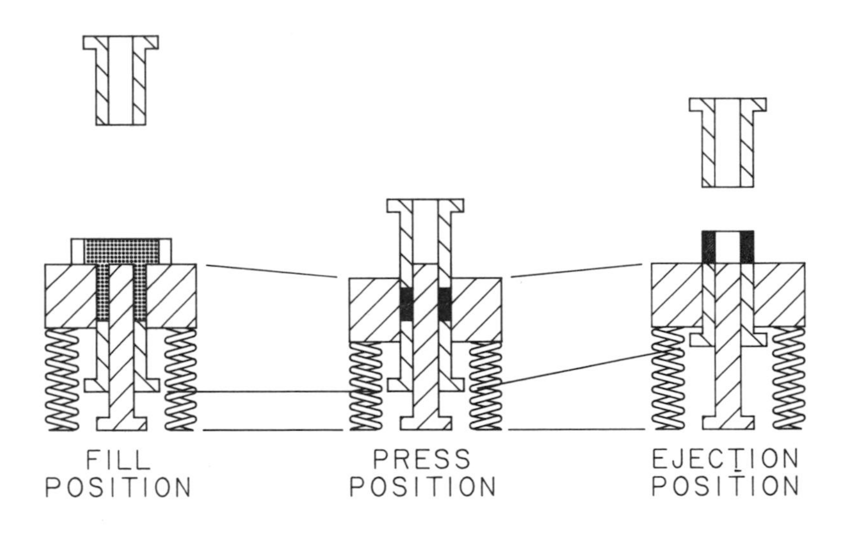

FIGURE 21-12

Upper punch pressing-floating die.

C. Upper punch pressing with floating withdrawal die: The floating withdrawal die system uses a free float of the die table to the compression position. The compacting phase is very similar to the system described in B above. Parts are ejected by the downward movement of the die element or table over the fixed lower punch. The fixed lower punch establishes a reference level for the die table in the fill, press, and ejection positions. Most systems of this type have mechanical stops to assure accurate positioning of the die with respect to the fixed punch at the compacted position (fig. 21-13).

D. Upper punch pressing with controlled withdrawal die: In this system the lower punch remains stationary as the dimensional reference point for adjusting the die table positions for fill, compression, and ejection (fig. 21-14). Adjustment of timing of the die travel provides positive control over the position of the part's neutral axis.

To generate the non-simultaneous double action movement:

(1) the upper punch moves down and seals the die cavity. The upper punch and die now travel at the same speed, relative to the fixed lower punch, until the die controlling element reaches a positive mechanical stop. This movement sets up single action compacting over the fixed punch and will place the neutral axis at this time at a point slightly above the center of the part.

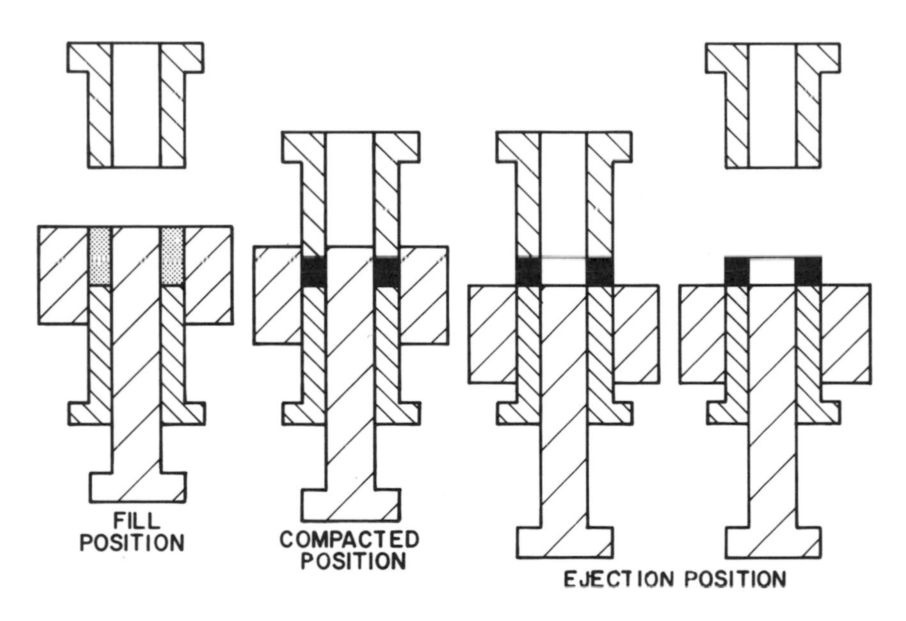

FIGURE 21-13

Upper punch pressing-floating withdrawal die.

(2) prior to completion of compacting, the die is held stationary against its stop, while the upper punch penetrates the die further. This final movement places the neutral axis back in the center of the part. Figure 21-15 illustrates the positive control of the neutral axis by the final upper punch penetration or pre-press adjustment.

A core rod is shown in fig. 21-14 attached to and moving with the die. As described in the section covering double action floating die presses (B above), the core rod can be arranged independent from the die or other tooling members. During ejection, in this case, the core rod is held stationary until final ejection whereupon it moves downward to a position flush with the die.

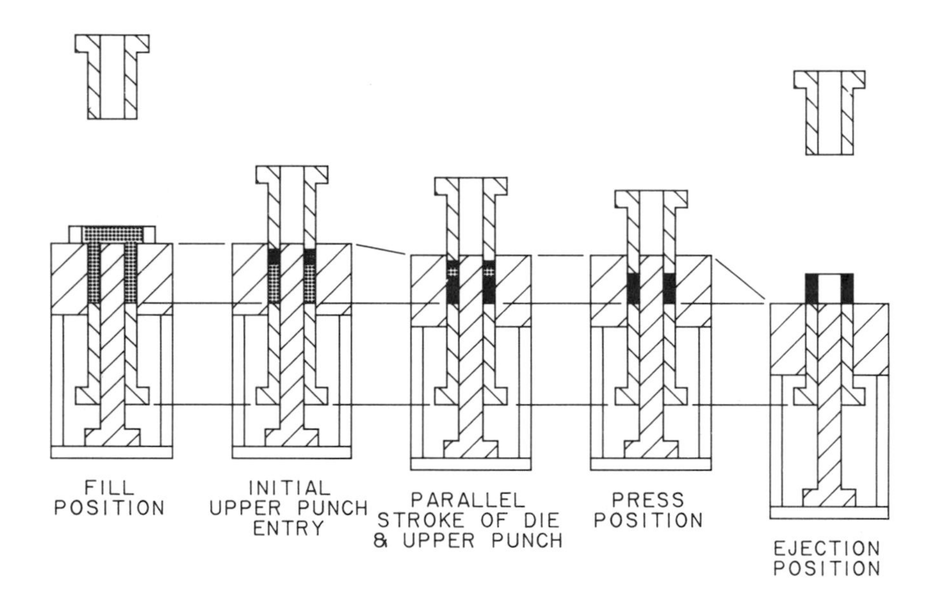

FIGURE 21-14

Upper punch pressing-controlled withdrawal die.

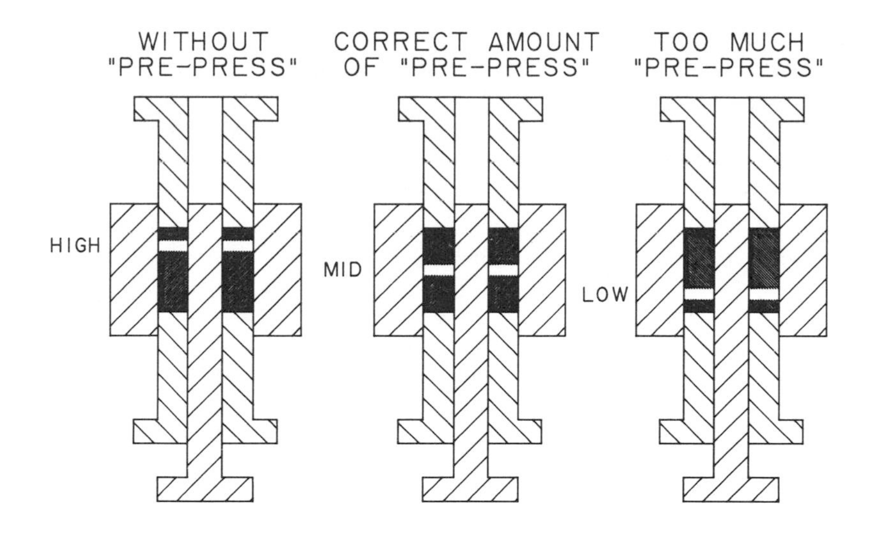

FIGURE 21-15
Neutral axis control

21.23 Multiple Motion Pressing

Multiple action presses, or tooling systems, are those which, on multi-level parts, support each level of the part with a separate punch or tool member. This system assures minimum density gradients in all levels of the part.

For example, if an external flanged part were compacted in a die with a fixed shoulder, the flange and body would have different green densities. In effect this would be single motion pressing on the flange and double motion pressing on the body. Usually the result is a crack on the underside of the flange where it joins the body. A further fault is that during sintering there would be differential movement (growth or shrinkage) between the flange and body, leading to warpage and loss of dimensional control. At ejection, regardless of the type of ejection motion (die withdrawal or punch pushout), the part is unsupported on the flange section and further cracking can develop. The same motions to form an internal flange or counterbore with a stationary type shouldered core rod would produce the same results.

Multiple motion compacting is usually required to press Class IV parts. Class I, II, and III parts can also be made in multiple motion compacting presses but, of course, with simpler tooling.

A. Multiple upper and lower punch pressing: The multiple motion press or tool system is similar in action to the double motion compacting system. On mechanical presses, the main punch motions are actuated by cams through linkages to the primary lower and secondary lower punches. Additional tool elements can be actuated mechanically, hydraulically or pneumatically. On all systems of this type, the die is held in a fixed position. Each lower part level is tooled with a separate punch. The distance at which the lower punches move relative to one another is normally adjustable, allowing a proportionally uniform buildup of density to take place for each level of the part. The tooling members normally have independent adjustments for powder fill, compression, and ejection. To produce double hub parts, the timing of the secondary lower punch motions can be adjusted to transfer powder into the different upper punch levels. Ejection takes place by upward movement of the lower punches and is sequenced so that the part is completely supported throughout the ejection cycle.

Figure 21-16 illustrates a typical production cycle for a Class IV shape. The configuration requires powder to be transferred from the lower punches into the upper punch hub area before compaction begins. Powder is transferred by adjusting the upper inner and outer punches to provide for an adequate volume of powder to form the proper density in the hub section. The upper outer punch will enter the die ahead of the upper inner punch, and this will result in a loss of powder unless provisions are made to confine the powder within the designated volumetric cavities. Usually this is done by positioning a moving core pin (mechanical, pneumatic, or hydraulic) in the upper inner punch to seal off the center section of the part for transfer.

After powder transfer, final compacting takes place by all lower punches moving up and all upper punches moving down. The center core in the upper punch floats upward within the upper punch and the die remains stationary. Ejection occurs as all lower punches move up, supporting each level of the part until each punch face is level with the die surface.

The core rod can be held stationary, as shown in fig. 21-16, to strip the part from the core rod as the part is being ejected, or it can be allowed to move up with the part during ejection until the part is fully ejected. The part then expands slightly allowing the core rod to move down freely to its fill position.

B. Upper punches pressing with floating withdrawal die: In the floating die multiple-motion press the lower motions are obtained by floating or counterbalancing the die and lower punches with air, oil or spring cushions. Punches are usually mounted on individual platens or rams which have independent adjustments for fill, press and ejection positions. The punch forming the lowest surface of the part does not move and is considered as a stationary ejection reference. The die platen must be driven down to a point flush with the top surface of this punch for full part ejection.

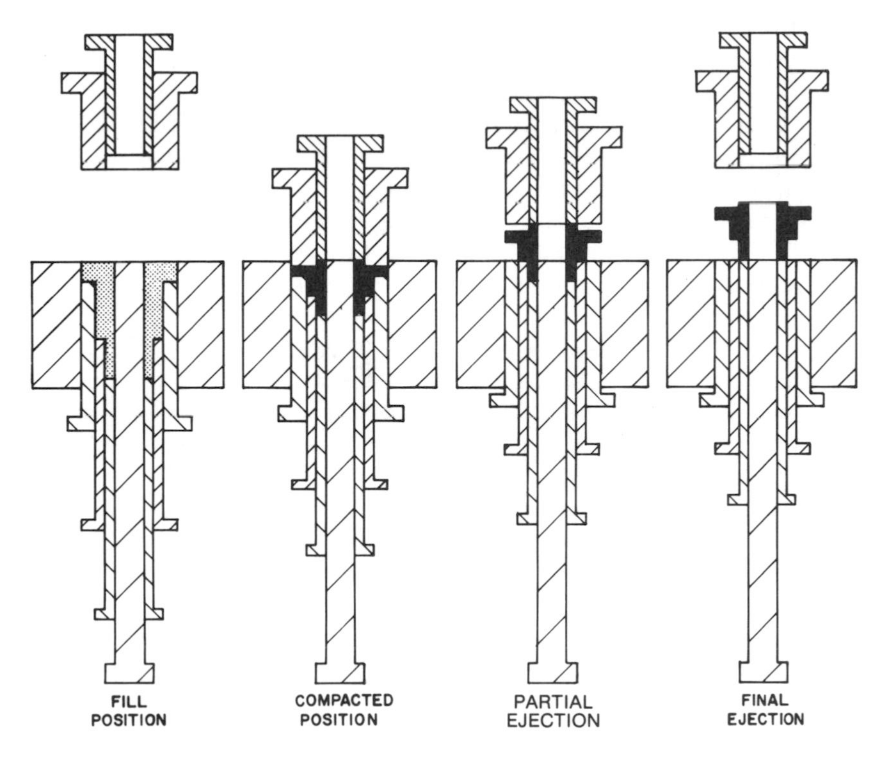

FIGURE 21-16
Upper and lower punches pressing-die stationary.

Figure 21-17 shows a typical Class IV part as compacted by such a system. All moving tool elements travel downward to their positive stops during the compaction portion of the cycle. Before ejection, these stops are released to allow the lower platens or rams to move further down. By timing the die descent in relation to the upper punch movement and through control of the counterbalancing forces on the tooling platens or rams the position of the neutral zone of the part can be controlled. During ejection the upper punches move away from the part, while the lower punches, move downward in sequence until the die and all the lower tool members are level with the top of the stationary or fixed punch.

The core rod can be provided with a pressing position stop to allow a part to be produced with blind or counterbored holes.

During ejection all tooling members except the stationary lower punch move downward with respect to the part. The core rod can be held stationary until the part is free of all other tooling members. The core rod is then free to move downward to the final ejection position.

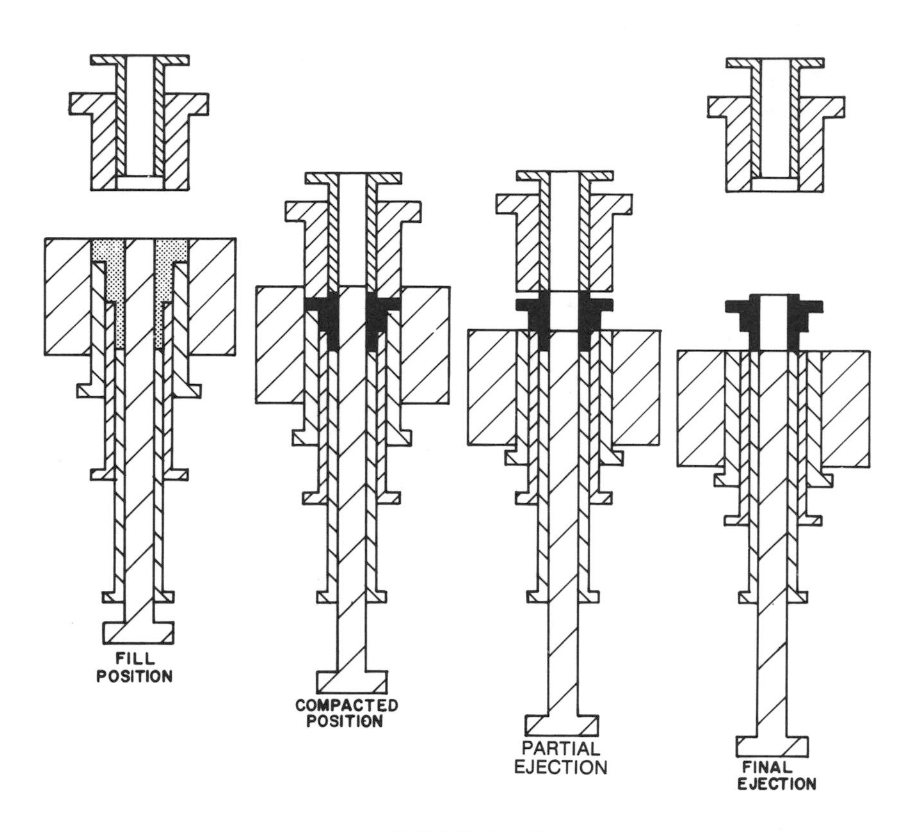

FIGURE 21-17
Upper punches pressing-floating withdrawal die.

At this point in the cycle the feeder moves across the die, pushing the compacted part from the die area. With the feeder over the lower tooling, the die and outer lower punches move upward to their fill positions. Next, the core rod moves upward, displacing the excess powder upward into the feed shoe. The feeder retracts, wiping the top surface level, readying the press for the next compacting cycle.

C. Upper punches pressing with controlled withdrawal lower punches and die: In the controlled die movement multiple motion withdrawal system, all moving tooling elements travel downward through the compaction cycle as with the floating die system. However, this system utilizes the removable tool holder or die such as shown in figure 21-18. Auxiliary punches are mounted in the die set with independent controls for fill, press and ejection positions. The compression stops for the secondary punches are mounted on slide blocks on the stationary platen of the die set, and are moved out of the way for ejection. The correct fill for each section of the multi-level part is achieved by first raising the die for the proper fill ratio of the section of the part above the fixed or stationary punch, and then by independent adjustment of the secondary or auxiliary punches for their respective fill positions.

FIGURE 21-18
Removable tool holder.

Figure 21-19 illustrates a Class IV part compacted by this method. During compaction the upper punch enters the die cavity and the die descends, escorting the secondary lower punches to their respective positive stops at final compaction. Powder is transferred into the secondary upper punch by proper timing of the die movement. There must be an upper core member, as in the multiple motion opposed action system, to assure powder confinement during transfer. Ejection is by the continued downward motion of the die. The positive punch stops slide out, so that all levels of the part are fully supported until each level is free of the die cavity. The core rod cycle is the same as described under 21.23 B above.

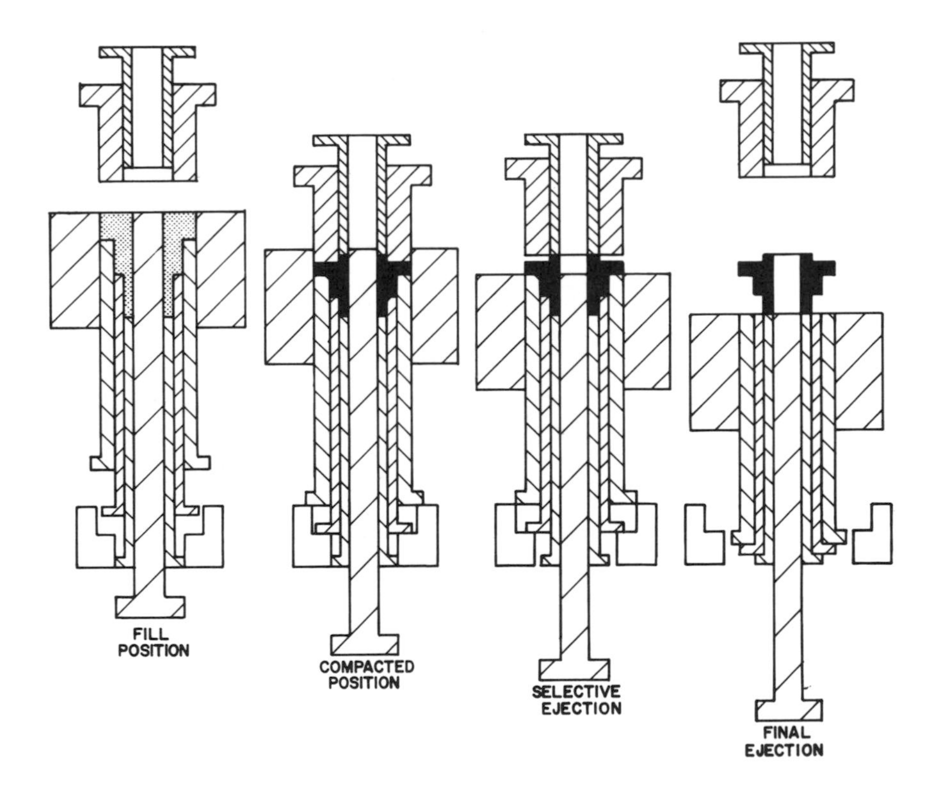

FIGURE 21-19

Upper punches pressing-controlled withdrawal die and lower punches.

21.30 Tooling and Tool Design

Every part being considered for powder metallurgy must be analyzed thoroughly. Powder specifications and pressure requirements must be determined in order to fit the capabilities of the equipment. The motions necessary to produce the part must be determined to utilize most effectively those that can be furnished by the press and those that must be incorporated in the tooling. Good tooling design is a key element in making a P/M part at minimum expense.

The following section on tooling and tool design for powder metallurgy contains basic information to assist the engineer in designing sound tooling and to outline for the purchaser of powder metallurgy parts the design parameters so he can gain from it a better understanding of the capabilities and limitations of the P/M process.

Figure 21-20 illustrates and names most of the normally used P/M tooling members. Figure 21-21 illustrates a tool set for making a gun part shown near the lower left.

21.31 Die

Most presses have large die openings to provide ample space for adapting tools from one make or size of press to another, and to allow room for shrink rings when insert type dies are to be used.

When calculating the necessary die wall thickness, one assumes that full hydraulic pressure is transmitted to the die wall. It is also assumed that the tensile stress in the die wall is distributed over an area three times the thickness of the piece.

Die inserts may be held in place by clamping or by shrink fit. Die bodies for shrink fits are preferably made of steels which are not as hard as, but much tougher than the usual die steels. Chrome nickel tool steels are commonly used. The tougher die steel is also used to minimize thread failure when the die insert is held in place by screws. When designing shrink fits, the usual interference for steel inserts is 0.2% of the insert diameter. It is recommended that as large a diameter as possible be used to support the insert and to prevent it from "breathing" and eventually failing. Carbide inserts are shrink fitted to only aout 0.1% of diameter of the insert since carbide is more rigid than steel.

Shrink ring type dies should be designed so compacting stresses never cause the insert or die liner to be stressed in tension. The shrink ring applies a compressive stress on the die inserts. During compacting this stress will fluctuate but should always remain in compression. Care must be taken to allow for stress concentrations at sharp corners or large changes in cross section of the die insert such as shown in figure 21-22.

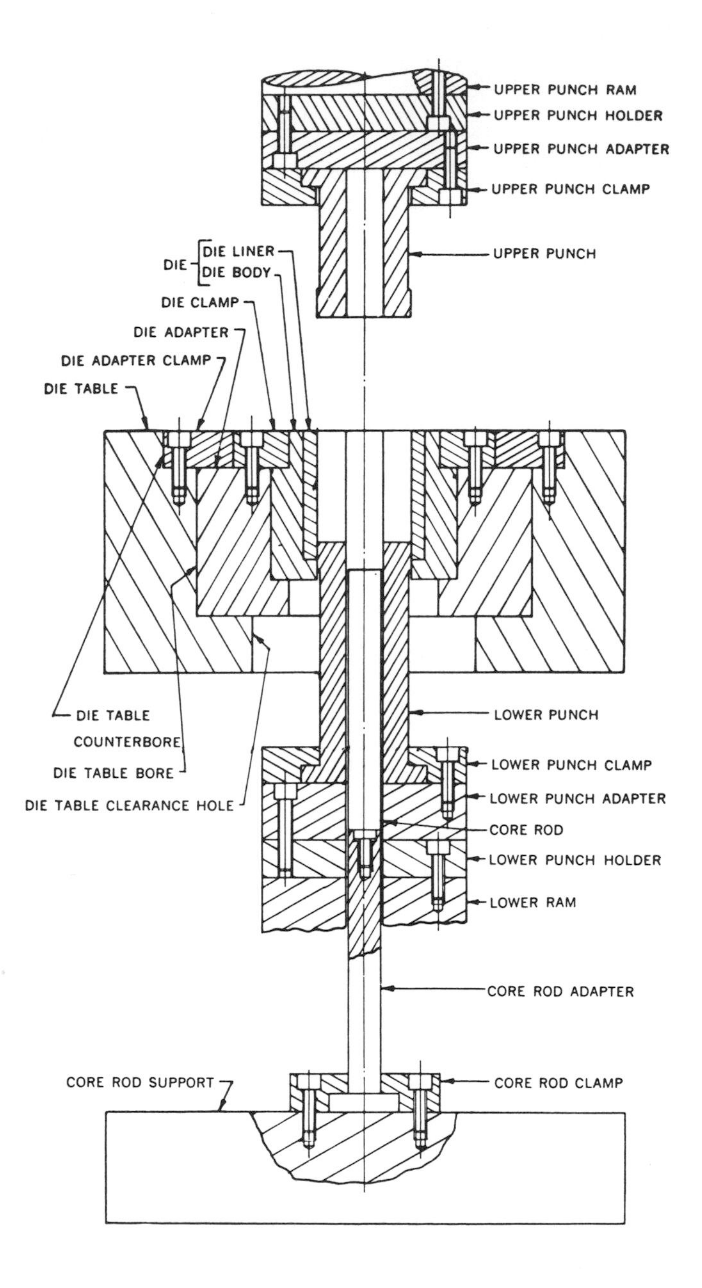

FIGURE 21-20

Cross section of typical P/M tooling.

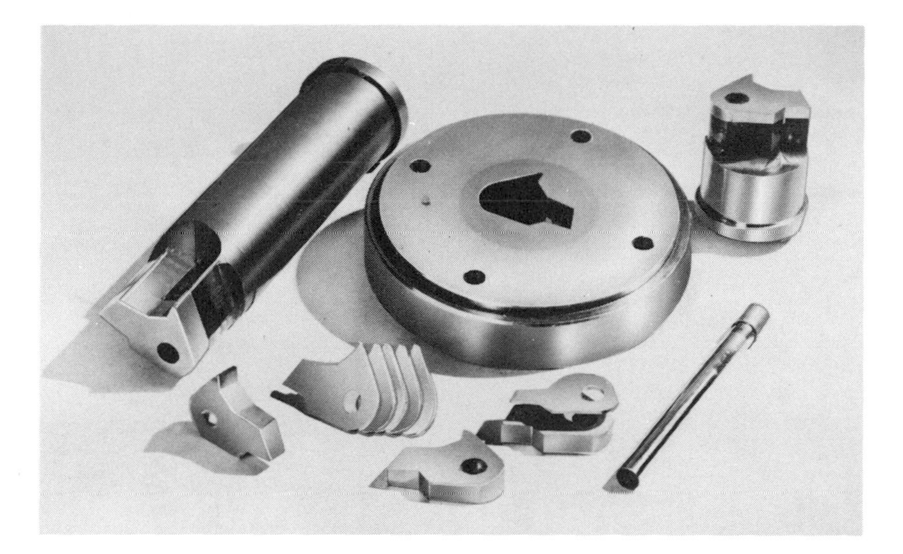

FIGURE 21-21
Tool set for making gun part shown at lower left.

The stresses in the shrink ring will be in tension. These stresses will fluctuate from the tension stress due to the shrink fit to a higher tensile stress at each compacting cycle, subjecting the shrink ring to fatigue stress. Most engineering textbooks provide formulae for calculating stresses for shrink ring construction.

Die entrance edges should be beveled about 15° from the vertical or radiused. The bevel should be 1/32 in (0.8 mm) deep, or less, depending on the size and the thickness of the pressed pieces. This bevel minimizes damage to the punch face edges when setting up and operating. Horizontal joints in or near the pressing area of the die wall should be avoided since fine powder can enter such joints and spread the sections vertically in spite of all precautions regarding finish of mating surfaces and high retaining pressures.

It is sometimes necessary to taper dies to aid in relieving part expansion strains during ejection which may lead to laminations in the ejected parts. Horizontal laminations or expansion cracks in the P/M part are most often caused by dies with insufficient stiffness to resist the radial forces properly during compaction. Tapering of the die increases difficulties in aligning the tools during setup. When a taper is required it is usually not necessary to make full allowance for the complete expansion of the piece. An allowance of 2/3 of the expansion is usually sufficient. If, for example, the pressed piece after ejection is 0.006 in (0.15 mm) larger than the die at the compression point, the taper can be 0.004 in (0.10 mm).

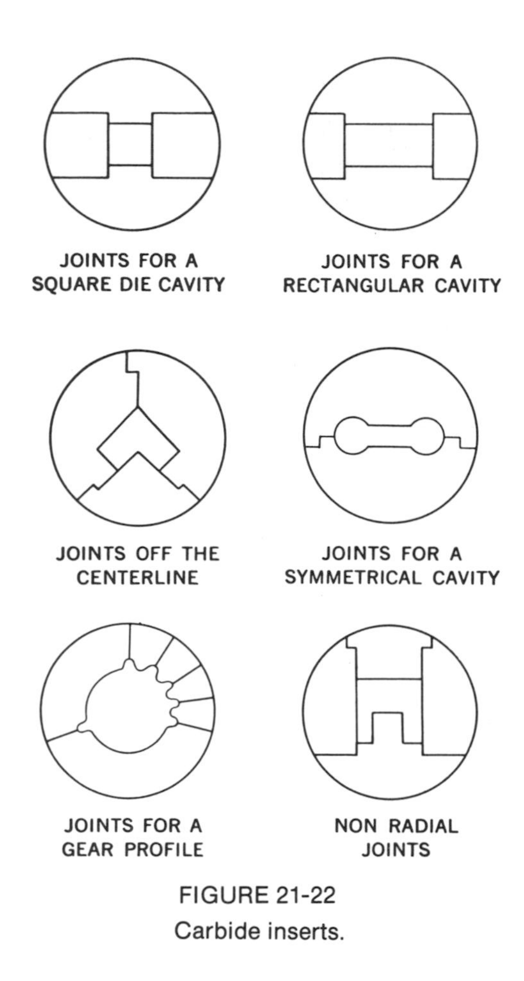

FIGURE 21-22
Carbide inserts.

The useful life of a die depends on many factors, such as the nature of the material being pressed, the unit pressure to be used, the allowable tolerances in the finished compacts, the material of construction and the surface finish in the cavity. High-chrome, high-carbon steels are used for medium production runs. The composition of such steels is usually about 12% chromium and 2% carbon. Both oil hardening and air hardening grades are available. The air hardening type is used for dies having sharp cornered cavities which might not withstand the shock of oil quenching. These dies should be heat treated to a hardness of 60-64 Rockwell C. When production quantities are medium to high, and/or increased wear resistance is desired, high speed steels such as AISI M4 are used.

When production volume is high or abrasive conditions are anticipated, dies should be made of tungsten carbide. Table 21-II lists various cemented

TABLE 21-II
Cemented Carbide for Powder Metallurgy Tooling Properties and Typical Applications

NO.	%. OF BINDER	HARDNESS, RA	TRANSVERSE RUPTURE, PSI	COMPRESSIVE STRENGTH, PSI		TOOL APPLICATIONS		
						CORES	PUNCHES	DIES
C-4	3%	92.3	177,000	800,000				Bearing Dies
C-9	6%	91.5	230,000	710,000	↑ SHOCK RESISTANT	Simple Shapes Short Lengths		Straight Thru Dies—Simple Cavity Contour
C-10	6 to 9%	90.6	280,000	650,000			Ceramics-Ferrites High Polish & No Face Projections	
C-11	12 to 13%	89.7	310,000	600,000		Step Cores & Complex Contours	Ceramics-Ferrites Metal Powders Simple Face Projections	
C-12	14 to 15%	88.5	340,000	580,000		Step Cores & Vulnerable Contour		Complex Shapes Gear Forms Sectioned Dies
C-13	15 to 20%	87.4	375,000	550,000	↓ WEAR RESISTANT	All Cores within Physical Limits of Carbide	All within Physical Limits of Carbide	Multi-Level Dies Vulnerable Projections
C-14	20 to 30%	82 to 86	365,000	470,000				

All property data represents average for grade

carbide grades and indicates which are best suited for simple or complex core rods, punches and dies.

For making unusual shapes a number of carbide inserts are usually fitted together and held with a shrink ring as shown in figure 21-22. This construction minimizes costly machining operations of solid carbide dies.

21.32 Punches and Core Rods

Toughness is important for punch steels. High carbon, high chrome steels are too brittle in most cases but 3% nickel, 1.5% chrome and 0.4 to 0.5% carbon steel is normally used, with the lower carbon preferred when sections are thin or chamfered edges are present. Special analyses in the AISI 3400 class meet the above specifications. For less delicate parts, AISI 320 class steel containing 1.75% nickel may be used.

When abrasion of the punch face is severe, punch inserts in chrome nickel holders are used. The inserts can be made of high-chrome steel with 1.5% carbon instead of the 2% carbon steel used for simple shapes. If bevels and other stress concentrating details are present, the 5% chrome-steel can be used. In cases of multiple punch tooling where a punch may function partially as a die, the use of 1.5% carbon high chrome steel or the 5% chrome steel is advisable.

Carbide punches are sometimes used for pressing abrasive materials. The more durable 9% or 12% cobalt grades should be used rather than the brittle or hard grades usually used for dies. When using carbide tips, the back up steel should be low-chrome rather than high-chrome tool steel to minimize mushrooming of punches under pressure.

Punches and core rods are relieved 0.005 to 0.010 in (0.13 to 0.23 mm) on the diameter and 0.0025 to 0.005 in (0.06 to 0.13 mm) all around on all profiles to permit the escape of powder passing down beyond the punch faces. The close-fitting portions are made as short as is possible. Care must be taken, in considering the length of each fitting portion, to allow for relative motion between parts. Punches, particularly if they form chamfers, tend to chip at the edges, and may require regrinding several times in their useful lives. Some allowances must be made for this on the assumption that each regrind removes at least the length of the chamfer and part of the close fitting portion of the punch or core rod.

Special considerations must be given to punch flange design. If the punch adapter is supported in a manner that it will allow the adapter to deflect, or "Oil can," the compacting force will be supported by the punch flange edge (fig 21-23). Relief of the flange will alleviate this condition.

On presses utilizing a withdrawal floating die, multiple punch arrangement (fig. 21-17), the punches undergo a compressive load during compacting,

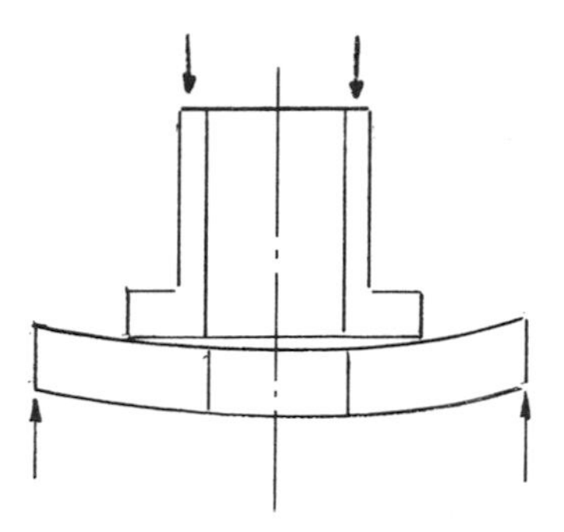
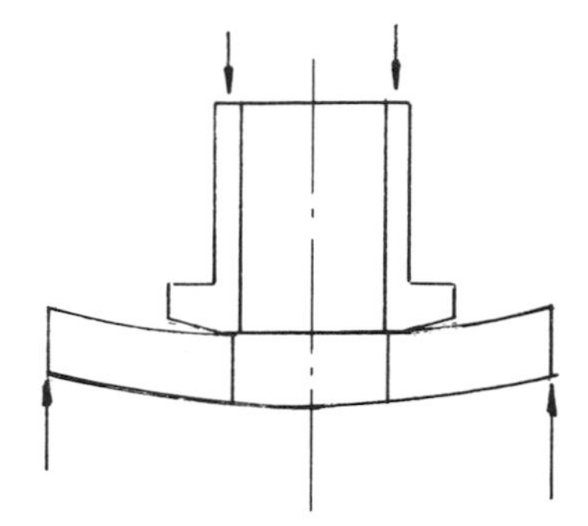

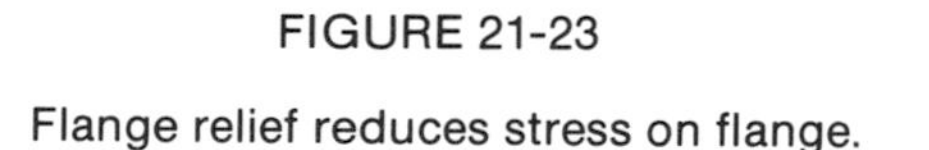

FIGURE 21-23

Flange relief reduces stress on flange.

with some punches being subjected to both compressive and tensile loads during ejection. The punch flange, clamp ring, and clamp ring bolts must be of sufficient size to resist the tensile load during ejection.

In multiple tooling arrangements, some punches must serve partially as a die; that is compacting takes place inside a punch. In such a case, the punch must be backed-up by the die for the full length of the compact. Punches subjected to such a load should be made of a material which is "tougher," than normal especially if the punch has an irregular sharp cornered gear tooth or contour.

Core rods are used to form through holes as well as blind or shoulder holes. When used to form through holes, wear resistance is of prime importance. High carbon-high chrome steels and high speed steels are commonly used. When core rods are used to form blind holes they are subjected to column loading during compacting and tool steels with additional toughness must be used in this application. The smaller diameter of the working length of a core rod should be made as short as possible.

21.33 Tolerances and Clearances

Punches should be made to fit the dies within a specified clearance rather than made to tolerances. The tolerances should be specified for the die and core rod dimensions, then the mating punches should be fitted to suitable clearances. The closest fits are required for making bearings and bushings. Any slight side-to-side variations in powder fill tends to push the core rod to one side which will take up all the clearance in one direction and produce eccentric bushings. No amount of subsequent pressure in the sizing operation can entirely correct the original eccentricity. For bushings the diametral clearances are usually not over 0.0002 in (0.005 mm). Eccentricity of ID and OD of punches should not be over 0.0002 in (0.005 mm) T.I.R. For other applications the clearances are generally more liberal 0.0005 to 0.001 in (0.013 to 0.025 mm) on diametral dimensions.

In making punches, concentricity of the punches with the punch shank or punch holder is not critical. Most presses include provisions for locating the punches concentric with the die so that alignment can be obtained. However, there is no provision for adjusting out-of-squareness so the squareness must be accurately maintained when making the tools.

21.34 Tooling Finishes

Die cavities and the core rods should be lapped or polished to a high finish after final grinding and the last polishing or lapping should be parallel to the axis of the tools. A five microinch finish or better should be achieved. When surface finish cannot be checked readily with a profilometer, a visual check for "mirror" finish by an experienced toolmaker is satisfactory. A well polished surface should have the same characteristics as a glass surface.

Punch faces and punch "lands" should have the same surface finish as the die cavities, and the final polish on the punch lands should be parallel to the punch axis. Punches and dies with poor surface finish wear out of tolerances much faster than properly finished ones, and may prevent the tooling elements from moving freely.

21.35 Punch and Die Adapters

Proper preliminary planning and ingenuity in adapter design can lead to large savings in tool cost. By the use of proper adapters, the basic tool element can be held to a minimum size. Where a variety of presses are in operation it is advisable to provide punch and die adapters so that tools from one press may be operated in other presses of different size or make. When designing adapters for complex multi-level parts, proper design of tooling adapters is a key element in producing high quality parts. The designer must consider stress and deflection, keeping the stress level well below the endurance limit of the material, yet balancing the deflection between the different levels of the part so as not to damage the part due to uneven deflection of the various tooling and pressing members.

Adapters should be made of steels having adequate strurctural and dimensional stability to maintain the accuracy of basic tool elements. Punch adapters should be heat-treated to a sufficient hardness to prevent "Brinelling" of punches into the adapter during compacting and ejection.

21.36 Tools for Coining or Sizing

Compacts coming from the sintering furnace may be off size. Such parts can be repressed, sized or coined to increase their density, to reshape them or to correct dimensional variations. Repressing can be done in the pressing die or slightly different die in the same press, equipped with a part feeder instead of a powder feeder, or, for short runs the parts can be fed by hand.

In many cases the tools are very similar to the compacting tools and are made of similar or slightly harder materials. Sizing pressure may be the same as or up to twice the compacting pressure. The die and core rod are provided with a tapering lead-in to assist the entrance of the parts into the die.

The core pin in some cases is an integral part of the upper punch, or it may be a separate upper punch, or the hollow upper punch may be spring mounted on the core rod pin.

Another sizing method, used chiefly for self-lubricating bearings, is to force the bearing into the die and then run a spherical burnishing tool through to size and to refinish the inside diameter. Self-aligning bearings and other spherical parts require special treatment to remove the central flat left by the forming process. Half the spherical section of these parts is sized in the die and half in the upper punch, which comes down and meets the die but does not enter it. The upper punch has a thick wall to withstand the sizing pressures.

A water soluble lubricant is used in most restriking operations. The parts are generally hand dipped on hand fed operations. The parts or tooling are sprayed in automatically fed presses.

21.40 General Press Information

21.41 Types of Press Drives

Both mechanical and hydraulic drives are used to provide motion and force for the compacting tooling that forms P/M parts. Generally presses of either type are available for producing a given part. Pneumatic and/or hydraulic auxiliary systems may be used to accomplish or to control specific functions during the press cycle. Presses from 3 to 1000 tons (2.7 to 907t) compacting capacity are available. Press cycles are adjustable to operate at up to 30 parts per minute for the smaller mechanical presses and up to 15 parts per minute for the larger hydraulic presses. The optimum operating speed is dictated by the size and complexity of the P/M parts to be made.

Some of the major advantages of each type are listed below:

Mechanical Presses:

- High production rates
- Low horsepower requirements
- Large range of pressing forces

Hydraulic Presses:

- Overload protection
- Greater depth of fill
- Lower initial capital investment

A. Eccentric or Crank – The most common type of mechanical press is the eccentric or crank type which converts rotary motion to linear motion (fig. 21-24). The mechanism permits high loading with low torque requirements at the maximum compression point (bottom dead center) with a low final pressing speed. Pressure in this type of unit is normally applied from the top only with an adjustment of the eccentric cam or in the Pitman arm link, to control the amount of upper punch entry into the die.

B. Toggle – The second most common type of mechanical press drive is the toggle or knuckle type. Actuation is usually accomplished by an eccentric or crank which straightens a jointed arm or lever, the upper end of which is fixed at the top while the lower end is guided for controlled accurate punch guidance into the die. The toggle design enables the press to develop very high pressure with low power and to provide minimum pressing speed near the end of the compacting stroke. This provides, in effect, a dwell at the compression point of the press cycle (fig. 21-25).

C. Cam – The third type of mechanical drive is the cam type. These presses utilize cam and lever arrangements to convert rotary motion to linear motion. Pressing speed, timing, and motion are controlled by changing the contours of the cams or cam inserts (figure 21-26).

D. Rotary – This machine has a series of punches and dies arranged in a common tool holding member, the press head or turret, which rotates around a spindle or stand providing a fixed reference point for mounting the press cams and pressure rolls. The cams provide vertical motion to the top and bottom punches. Rotary presses are best suited for smaller, high volume, high production applications. Production rates can range up to 1,000 pieces per minute depending upon the material flow, part configuration, part dimension, and pressing tonnage (figure 21-27).

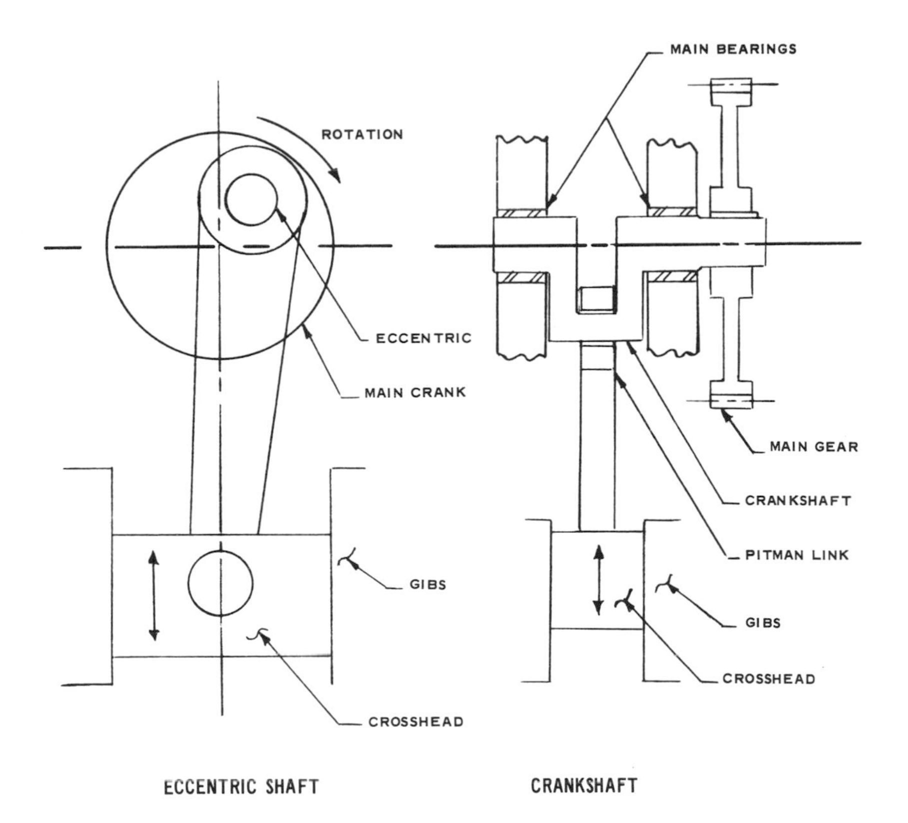

FIGURE 21-24

Eccentric or crank drive.

E. Hydraulic presses depend on fluid pressure, acting on one or more rams to drive the upper or lower punches, or both. The timing of the punch strokes is usually controlled electrically. Hydraulic press capacities range from 100 to 3000 tons (90 to 2720t).

21.42 Powder Feeding

Automatic powder feeding and part removal are achieved normally by the press powder feeding system. In general, it is desirable that the feeder cover the die opening while the bottom punch is at the ejection position. This minimizes the chance of air entrapment, which could hinder consistent powder feeding. If the bottom punch has a projection on it, it is impossible to completely cover the die with the feed shoe while the lower punch or die are at

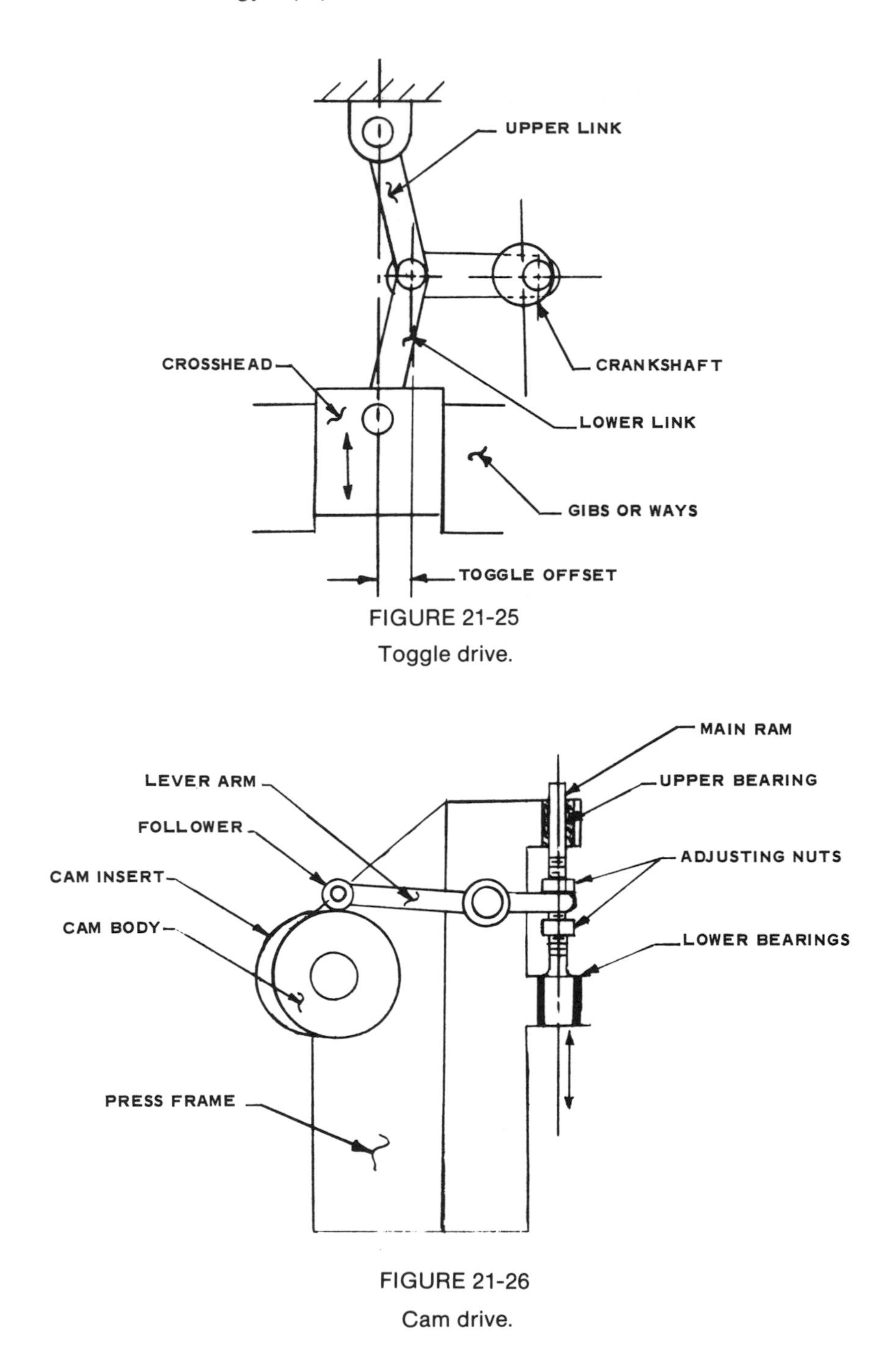

FIGURE 21-25
Toggle drive.

FIGURE 21-26
Cam drive.

ejection. However, it is still desirable to time the feeder stroke to cover the die as soon as it can clear any lower punch projection. Most press manufacturers can supply auxiliary feed shoe agitation attachments to aid filling of material under difficult conditions. There are three basic types of feeders used on compacting presses.

One type, a direct shuttle feed normally used on moving die tables, provides a straight in-line reciprocating action over the die with the feed shoe connected directly to the supply hopper which pivots on its mounting. The feed shoe motion is imparted by a direct connection of the shoe with a mechanical, penumatic, or hydraulic system (figure 21-28). Additional agitation of the shoe can be provided by contouring the feeder cam in a mechanical system or by pulsating the pneumatic or hydraulic system.

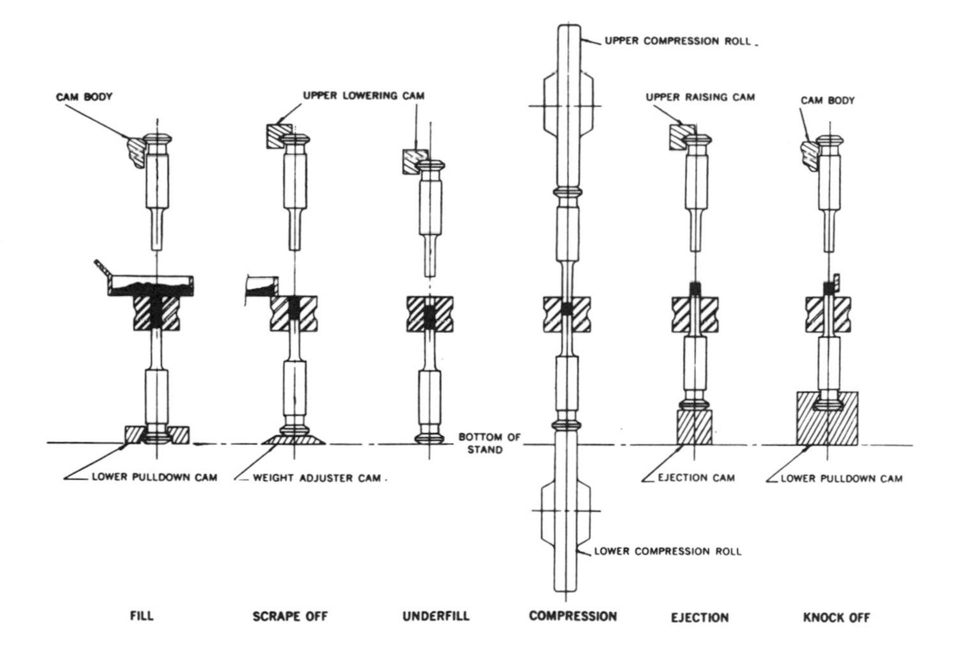

FIGURE 21-27
Rotary drive.

Another type is a metered shuttle feeder. Its motion is the same as the in-line system. However, it does not have a direct connection with the supply hopper. At fill it moves with a metered amount of material from a position under the hopper to a position over the die cavity. At the same time it closes off the powder coming from the supply hopper with a cut-off plate. Its motion is imparted by a direct connection of the shoe with a mechanical, pneumatic, or hydralic force system (figure 21-29). This system supplies the same metered amount of material over the die cavity on each press stroke.

A third feed system, which is normally supplied on mechanical presses with a stationary die table, uses a pivoting action of the feed shoe over the die area. This system is commonly called the arc type feeder. The shoe is pinned or guided on the die table while its movement is controlled by a feed actuating cam (figure 21-30). The hopper is mounted directly above the feed shoe, providing a constant pressure head of feed material into the shoe. Agitation of the shoe is usually accomplished by contouring the feeder actuating cam.

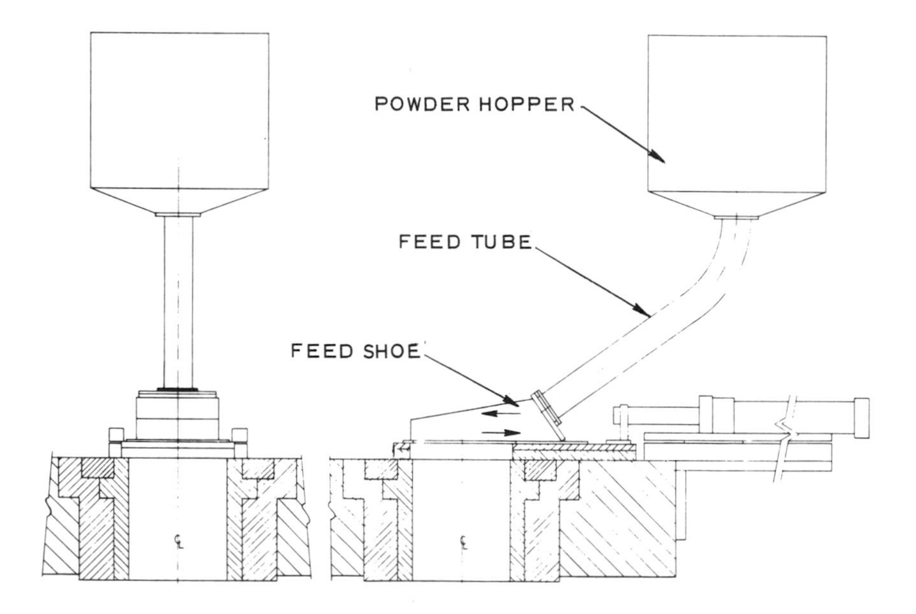

FIGURE 21-28

Direct shuttle feed.

21.43 Optional Attachments

Most press manufacturers offer a number of press attachments which can be subdivided into two major categories.

A. Control and maintenance options: Air clutches, specialized electrical control systems, lubrication systems and carbide wear parts in critical areas all serve to complement the basic press.

The air clutch, when offered on a mechanical press as an option, provides additional safety and reliability when there are repeated starting and stopping operations as are used for coining and sizing.

The installation of carbide buttons or wear plates in the gib-crosshead area on either mechanical or hydraulic presses will help maintain tool alignment and prolong the working life of the press. Their selection and use depends primarily on in-house operating conditions.

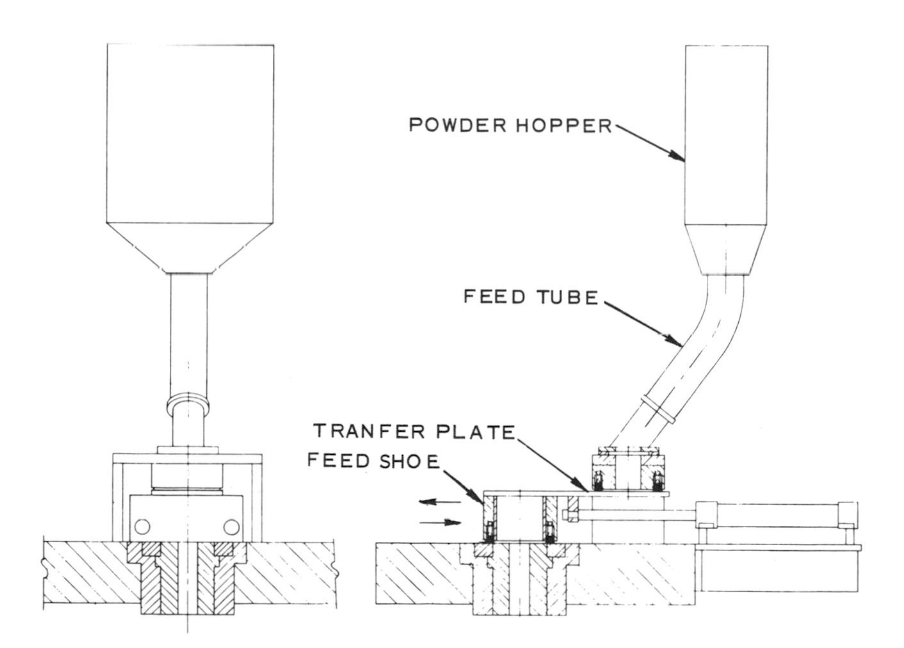

FIGURE 21-29

Metered shuttle feed.

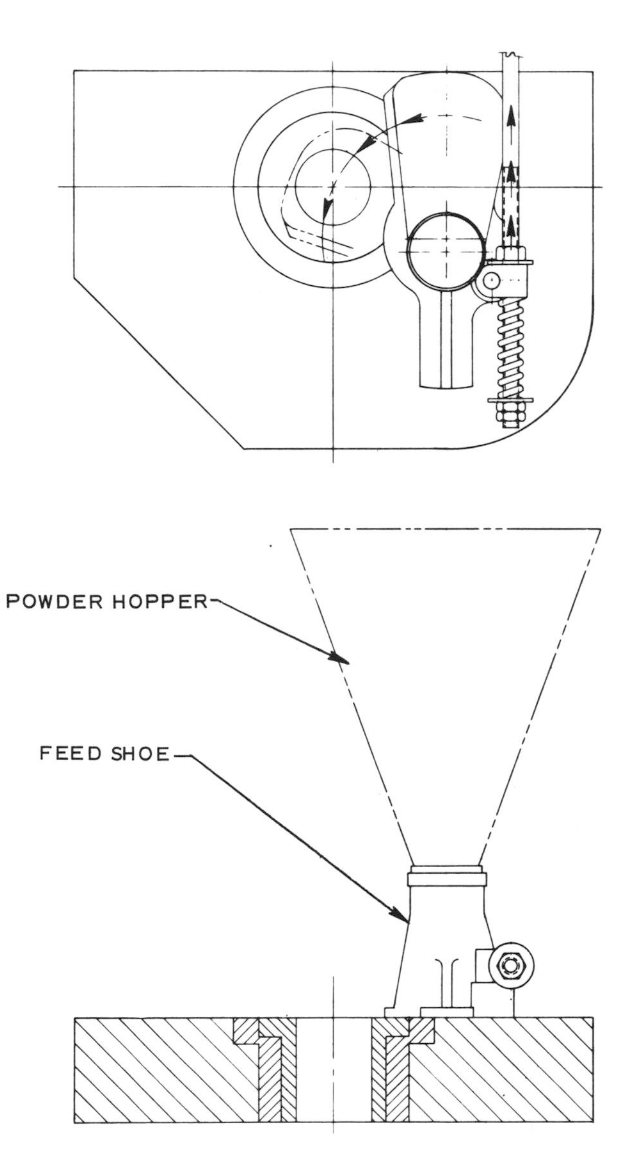

FIGURE 21-30

Arc type feed.

Automatic conveyor belts or part take-off devices, when offered for part removal from the press, should be considered seriously in terms of added safety and efficiency. Motorized upper ram, raising and lowering systems on large tonnage presses should be considered as a definite convenience and an aid in reducing tool set-up times.

Manual lubrication systems can be individual-point or centralized-point systems. The individual-point system requires servicing by a handgun at each point of lubrication. The centralized-point system employs a pump to force feed the lubricant to the individual points throughout the press. The pump is manually operated and may or may not be attached to the press. Individual-point systems use grease almost exclusively as the lubricant. The centralized-point system can be set up to operate with oil or grease.

Automatic lubrication systems are centralized-point systems equipped with a pump, programmed to force-feed the lubricant automatically to the Individual points throughout the press. The pump can be driven by an electric motor, pneumatically or driven from some moving member of the press. Automatic lubrication systems should be equipped with electrical interlocks that prevent the machine from running in the event of a lubrication system failure and/or low lubricant level in the reservoir.

Information on the frequency of lubrication and the type of lubricant can be obtained from the machine manufacturer.

B. Supplemental press and tooling options: Upper Punch Hold Down – This attachment consists of pneumatic, hydraulic, or mechanical device built into the upper press ram or one of the main upper compression members, to maintain some top pressure on the part during ejection. Top pressure at this point in the press cycle effectively increases the top punch dwell on the part and allows more time for entrapped air in the green part to escape. When compacting low green strength materials, such as ferrites and carbides which tend to laminate if air becomes entrapped in the compact, the hold down system supports the part during ejection while allowing entrapped air to escape.

Hydraulic or Pneumatic Equalizer Systems – These attachments are offered for mechanical presses to ensure the production of compacts of even density despite variations caused by irregularities in fill of material being compacted. They normally consist of a hydralic or pneumatic link in one of the main compression members of the press. The system is preset to the desired compacting pressure for the compact. If this pressure is exceeded, the link deflects while maintaining the desired, preset maximum pressure.

The system can also be used as an overload release by setting the pressure in the system slightly higher, than the pressure necessary to compact the part. For this application, the release mechanism is usually linked electrically with the press clutch mechanism. As the preset pressure is reached, the clutch is de-energized and the brake is applied to stop press motion.

Movable Core Rod – This option is offered for both hydraulic and mechanical presses. Motion of the core rod can be from mechanical, hydraulic, or pneumatic energy sources. The movable core rod system reduces ejection loads, aids in filling, or provides the relative motion necessary for compaction of complicated shapes. It also helps to hold dimensional tolerances of parts with long, small diameter holes.

The movable core rod mechanism is normally mounted on the lower portion of the press frame. The simpler types usually provide a standard two position adjustable stop motion in one of the three operations.

1. Filling and stripping core rod – In this operation the core rod descends to take on added material in the die cavity before the lower punch reaches the fill position. It is then raised prior to the completion of fill so the excess material can be pushed back into the feeder shoe. During compression the core rod remains stationary. During ejection it moves toward its lower stop. This action results in stripping the core rod from the piece, and places the core rod in the initial overfill position. On presses using the lower punch for ejection, the stripping core rod moves down before the lower punch moves up. This serves to decrease some of the ejection load on the press.

The filling core rod system can be used also to transfer powder into top hub part shapes when dual upper punches are utilized. In this case the transfer time is delayed until the upper punch starts its initial entry into the die.

2. Ejecting core rod: The top surface of the ejecting core rod is held flush with the top surface of the die during the fill and compression portions of the cycle. At ejection, it rises to its upper stop limit with the compact above the die table. The part expands away from the core rod and the rod is retracted to its lower limit. This action results in the decrease of ejection loading on the press and less wear on the core rod.

3. Floating core rod: The movable core rod during this operation is preset to a fill position and supported in this position by mechanical, pneumatic, or hydraulic pressure. During compression, the core rod floats down to a mechanical stop as the piece is being compacted. At ejection the core rod is returned to its original fill position. The floating core rod, when applied to a stepped core rod, provides relative motion between the lower punch and core rod. This helps control density variations in the part and eliminate shear planes and crack formations. Pneumatic and hydraulic floating core rods have the added versatility of adjustable resistance.

When applied to thin, long core rods, the core rod may float during pressing. This helps reduce the compressive stress and bending of the core rod and consequently the distortion of the hole.

On presses 100 tons (91t) and over, more sophisticated three and four position core rod systems are offered. Basically they provide a combination of the motions previously described. The four position core rod has two separate stop systems. The core rod can be set as a filling member for making top hub parts, with its adjustable stop locked in position for compression. After compression the stop is unlocked allowing the core to serve as a stripping tool during ejection of the part.

4. Overfill-Underfill: This option is generally available for either hydraulic or mechanical presses.

Overfill: This option provides an adjustable stop to allow an excess amount of powder fill over the lower punch at the first part of the fill cycle. Prior to completion of fill and while the feeder is still over the die cavity, the overfill mechanism is released so the die or lower punch can return to the correct fill position. This addition and removal of extra material improves powder-fill characteristics for hard-to-fill high length-to-diameter ratio parts and fine tooth gears.

Underfill: With this option, the powder level after filling has taken place, is lowered in the die by moving an adjustable stop controlling the die table or lower punch. This allows the upper punch to enter the die during the initial compression phase of the pressing cycle without displacing powder.

Dual upper punch system: This option, which can be mechanical, hydraulic, or pneumatic, is used when the part to be compacted requires forming an upper hub or a flange-type part. An inner upper punch and an outer punch are used to form the top of the hub and the top of the flange. Most systems provide adjustment for fill compensation on the outer lower punch, control of resistance to float-back of the outer upper punch during compaction, and an ejection motion behind the inner upper punch to strip the hub from the outer upper punch before the outer punch moves up.

21.44 Press Maintenance

Preventive Maintenance: A preventive maintenance program, if performed on a regular basis, pays for itself by uncovering minor problems which, if not found and corrected, could lead to major repair problems. The type and complexity of the press will determine the preventive maintenance program to be followed. The instruction manual supplied with the machine may outline the program. If not, the machine builder will help establish one.

A preventive maintenance program for mechanical presses should include daily, weekly, and monthly inspections and servicing of the lubrication points, inspection of cams and cam rollers, and a visual inspection of bolts and screws. The monthly inspection should include, in addition, a general cleaning of the press.

A program for hydraulic presses should include a daily inspection of the hydraulic oil level, temperature, and pressures. Provisions should be made for daily, weekly and montly inspection and servicing of the lubrication points. Weekly inspection should include cleaning of the hydraulic oil strainer, filters, and magnetic traps. Monthly inspection should also include a general cleaning of the press.

The user of a P/M press should establish and follow a program of regular inspection to ensure that all its parts, auxillary equipment, and safeguards are in proper operating condition. With the aid of the machinery builder the user should develop and maintain a checklist, and keep a record of all inspections and maintenance work performed.

Daily inspection of all interlocks and safeguards is recommended. An improperly operating device can be the cause of serious damage to the machine and/or tooling and may cause injury to the operator.

21.45 Press Safety

Safety is the joint responsibility of the machine manufacturer, the owner, the operator, and of the set-up and maintenance personnel.

The manufacturer's responsibility is to provide a machine which, when operated and maintained properly and used for its intended purposes, will not subject the machine operator to risk of injury. Instructions for normal safe operation and recommended maintenance are also part of the manufacturer's responsibility.

The owner's responsibility is to ensure that the operator, set-up and maintenance personnel follow the operating and maintenance instructions given by the machine manufacturer. The owner should be sure that all personnel observe and obey any warnings signs or labels that the manufacturer installs on the equipment. The owner must not alter, or allow his personnel to alter, the machine cycle, electrical, hydraulic, or pneumatic cirucits without first obtaining clearance from the machine manufacturer.

For more detailed information, the MPIF Standard #47 entitled "Safety Requirements for the Construction, Safeguarding, Care and Use of P/M Presses" can be obtained from the Metal Powder Industries Federation, 105 College Road East, Princeton, New Jersey 08540. The standard applies to mechanical and hydraulic compacting presses which are designed, modified, or converted to compress metallic and non-metallic powders. Additional information can be obtained from the National Safety Council, 475 N. Michigan Ave., Chicago, Illinois 60611.

State and local safety codes should be studied also to understand your additional responsibilities. Consultation with loss prevention departments of workers compensation carriers can also be helpful.

22.00 POWDER FORGING - P/F

The advent of powder forging (P/ F) has filled the gap in part properties between conventional cold pressed sintered parts and wrought parts. P/ F combines the advantages of the P/ M process with those of conventional forging and with dimensional tolerances usually in the range of ± 0.2%. In many cases P/ M allows the designer the flexibility to match the process to the property requirements of the part.

Potential advantages of P/ F over conventional forging are in material savings, reduced machining steps, fewer tooling steps resulting in less labor, less directionality of part properties, and a better surface finish. Usually a distinction is made between P/ F and hot sizing. Hot sizing, sometimes called hot restriking or hot densification, incorporates a minimal amount of material flow during the forming operation; P/ F refers to a process involving significant material flow.

The basic P/ F process involves the formation of a powder preform to precise weight tolerances, sintering the preform, usually in conventional sintering equipment, reheating the preform and a final precision P/ F operation.

22.11 Powders for Forging

Powder requirements differ for P/ F and P/ M parts. In the latter, finished parts have approximately 10 to 25% porosity which becomes the principle factor in determining the mechanical properties of the parts. As porosity decreases, other factors such as material impurities and inclusions, which were insignificant at the lower densities, become much more important in determining properties. These inclusions are unavoidable in steel making operations. Those present in powder are generally smaller and less deformed than the inclusions in wrought steel.

When considering the P/ F process it is necessary to evaluate the powder in terms of hardenability, oxide reduction, heat treatment, and machinability. If the hardenability is insufficient, the carbon content can be increased by making graphite additions to the powder.

22.12 Preform Design

The shape of the preform is determined by the final shape of the part and by the property requirements. The best combination of mechanical properties is achieved by maximizing deformation during forging. However, care must be taken to avoid cracks, tears or other flaws during deformation. Many P/ F preforms are now computer designed to obtain optimum deformation during forging.

Preforms can be made by any of the available compacting methods but it is important to hold close weight control. Because the P/F process occurs in a closed die, any variation in the preform weight will be reflected directly in the thickness of the final part. For example, an 0.005 oz (0.14 g) change in weight on a part with a 5 in^2 (3224 mm^2) cross section will change the height by 0.002 in (0.05 mm).

Preform density and density distribution become more important as the complexity of the part increases. On complex parts it is necessary to have the correct mass distribution in the preform to ensure proper material flow during forming. Improper mass distribution could result in low density in some areas and overloading of tooling in other areas.

Handling and transfer of a shaped preform require orientation prior to transfer to the forming die. For many parts a simple circular preform may be selected for ease of transfer.

22.13 Preform Sintering

Preforms are normally sintered in any of the conventional P/M furnaces. Although short time sintering by induction has been proven feasible, production equipment for this has not been readily available.

The oxygen level in the final part affects impact properties of the part but has a negligible effect on the tensile stength. Low oxygen levels can be obtained by sintering at a temperature of 2250F (1230C) rather than the usual 2050F (1120C) used for iron based materials. The sintering operation is also used to maintain the proper carbon level; control of carbon in the sintered preform is important to ensure the proper hardness and strength of the final part.

22.14 Preform Reheating

Prior to the forging operation, preforms are reheated to the proper temperature, usualy in the range of 1600 to 1800F (870 to 980C). Induction heating is the most common heating method used because of reduced floor space requirement and rapid heating capability. Sometimes, conventional furnaces are used. The disadvantages of induction heating center on difficulties in heating irregular shapes, and the changeover time involved when switching from one part shape to another. Before reheating, the preform is usually coated with a water-graphite solution. This solution inhibits oxidation and decarburization, avoids sticking between the preforms during heating, and provides die lubrication during forging.

Typical induction heating times are in the range of 45 to 90 seconds; for many parts, no atmosphere protection is required to prevent excessive oxidation and decarburization.

22.15 Preform Transfer and Forging

Various types of transfer mechanisms are used to transport the hot preform from the heating mechanism to the forming die. As a safety factor, these units are interlocked with the press either mechanically or electrically. A wide variety of forming presses, either mechanical or hydraulic, can be used. It is important to minimize contact time between the hot preform and the tooling to minimize heat transfer into the tools. When preform heating or handling is slow, the press should be run intermittently at its normal speed rather than running it at a speed to accommodate the slow heating or handling operations.

A typical sequence of motions necessary to forge a preform includes transfer of the hot preform into the forging die, cycle the press to form the part and eject it from the die, transfer the part away from the die, then cool and lubricate the tooling. In addition to the forging stroke, other motions usually required to form a part include lower punch ejection and an upper punch stripping motion relative to the ram movement. For some parts, such as those with a hub, it may be necessary to float the die by building this capability into the tooling or into the die set.

Preform temperature must be controlled accurately to ensure a consistent temperature during the forming cycle since temperature variations will affect final part dimensions. Improperly heated preforms during both production runs and start up or shut down operations should be rejected. Because temperature control is so important for dimensional stability, the tooling is preheated and maintained at a controlled temperature. To avoid heat build-up on the forming surfaces, the tooling is sprayed with a water-graphite spray mix after each forging stroke. This spray serves as a coolant and a lubricant.

22.16 P/F Tooling

Tool designs for powder forging are complex. The pressing forces required for P/F are approximately twice those for powder compacting. These forces combined with the greater transmission of lateral forces, result in radial pressure about five times that in cold compacting. Therefore, it is very important to determine the magnitude of shrink fit required to balance the stresses in the die insert and the shrink ring. If high ejection forces are anticipated, provisions should be made to avoid pushing the insert out of the shrink ring. A reverse taper is recommended rather than a step which would introduce a stress concentration factor. When choosing a material for the die insert, hot hardness, wear, and toughness are factors which must be considered and the relative importance of each may change for different parts. Even though the tooling is preheated to a uniform temperature, the surfaces in contact with the preform may fluctuate between 300 to 1200F (150 to 650C) and this must be considered when selecting the tool material.

Punches should be designed to be as simple as possible. Thin protrusions to form chamfers and steps to raise short hubs or counterbores are possible. However, such shapes introduce lateral forces in the punches and may make it difficult to remove the formed part from the punch. Also, the life of these tooling members will be shortened because of temperature fluctuations and movement of metal along the tool surface. Punches can be made of the same materials as the dies.

Core rods usually have a large area exposed to the hot preform and are difficult to cool. A P/F part contracts rapidly from the die wall and around the core rod after it is formed. It is easy to eject from the die, but difficult to strip from the core rod. The P/M compacting practice of waiting until the part is ejected from the die before stripping the core rod is not applicable in P/F. The core rod should remain in contact with the preform and part for the shortest time possible, and it should be the first of the tooling members to leave the part during ejection. For this reason P/F core rods are often incorporated with the upper punch.

Many factors must be considered when determining the room temperature size of the die cavity. These include the thermal expansion of the die due to preheating, the forming load, the elevated temperature yield strength of the part at the fully forged position in the die, and the thermal contraction of the hot formed part after it leaves the die. Using typical hot forming parameters, it would require a cold die cavity of 4.035 in (102.49 mm) diameter to make a part with a 4.000 in (101.6 mm) diameter. The most influential variable affecting die size is the part temperature at its fully forged position in the die. A decrease of 100F (56C) for the above example will change the required cold die cavity diameter to 4.030 in (102.36 mm).

Tooling clearances for hot forming are greater than those used for compacting tooling. A clearance of 0.001 in (0.025 mm) per side per inch of diameter can be used as a rule of thumb. Clearances of 0.002 to 0.002 in/in (0.05 to 0.08 mm/mm) are common on larger parts produced at high production rates.

22.17 Heat Treatment and Machinability

Because of the infancy of the powder forging as a production process, very little data have been collected concerning the heat treatment and machinability of P/F parts. The microstructures and properties will differ slightly from parts produced from conventional wrought steels since the powders have slightly different compositions. However, in general, standard heat treatments can be used for the P/F parts.

Most P/F parts require only minimal machining or grinding. Grinding wheel life tests have not been conducted, but the producers surveyed felt that there are no major differences in grindability between P/F and wrought forged steels. In cutting operations, the P/M hot formed parts may be more consistent in machinability than wrought parts because the finely dispersed inclusions in P/M parts yield short chips rather than long continuous chips sometimes encountered with wrought steels. For some machining operations, the parameters may have to be adjusted from those used on wrought parts. Often the overall machinability of P/F parts appears to be as good as or slightly better than that of equivalent wrought materials.

22.18 Hot Sizing

P/M hot sizing is very similar to the P/F process. The previous discussion on P/F equipment and tooling is applicable to hot sizing. The two processes differ in the magnitude of deformation and by how closely the preform resembles the finished part.

The hot sizing operation requires a preform near the final shape and dimension of the finished part. There is a minimum amount of metal flow. Tolerances and finishes are improved, but physical properties are not as good as those obtained by P/F. The force required is considerably less than that required for powder forging, and tool life is greater. For parts which require physical and/or mechanical properties intermediate between P/F and hot sized parts, the preform shape can be modified. Such preforms are designed to give only the amount and distribution of strength required. This technique minimizes the amount of press tonnage and die wear for medium strength parts.

23.00 ISOSTATIC PRESSING

Isostatic pressing is a P/M forming process that applies equal pressure in all directions (ISO) on a metal powder compact thus achieving maximum uniformity of grain structure and density. Simple or complex shapes requiring uniform strength and fracture resistance are the primary applications for the process. The process also lends itself to making larger P/M parts and those with extreme ratios of length to diameter.

While cold pressing has remained the dominant method of pressing metal powders because of the economy with which complex shapes of accurate dimensions can be formed, it has some technical limitations such as size or configuration that restrict its use. Isostatic pressing is usually adopted only when cold pressing is ruled out.

23.10 Cold Isostatic Pressing - CIP

There are applications in which cold isostatic compaction is preferred on economic grounds; notably, when producing a small number of parts where the high first cost of cold pressing tools cannot be justified, and when some machining is permissible, or when very large compacts are needed. The flexibility of equipment for wet bag compaction is another advantage; a single piece of equipment can handle one large compact or many small ones at the same time. Compacts of different sizes and materials may be processed together, provided the compacting pressure is acceptable for the shapes and materials involved. In the dry bag system the flexible molding bag remains in the pressure vessel and the pressurizing fluid is never seen. It is confined behind a sealing membrane. The vessel can be open top and bottom.

Pressure can also be varied easily from one run to the next. A variety of materials can be presed isostatically on a commercial scale, including metal powders, ceramics, plastics and composites. Pressures required for compacting range from less than 5,000 to 100,000 psi (34.5 to 690 N/mm^2) and above. Figure 23-1 shows schematic illustrations of the wet bag and dry bag isostatic pressing principles.

23.11 General Advantages of Isostatic Pressing

A. Uniform strength in all directions: A green compact produced isostatically has uniform strength in all directions, because the pressure used to compact the material is applied equally in all directions.

B. Uniform density: Pressure applied from all directions produces a uniformly dense part. The resulting piece will have uniform shrinkage during sintering with little or no warpage.

C. High L/D ratio: Long, narrow parts which are impractical to produce by conventional pressing, can be produced easily isostatically.

D. High green strength: Parts produced isostatically have relatively high green strength. This offers better possibilities for machining the compact in its green state.

E. Unlimited shapes: Isostatic pressing makes it practical to produce shapes and dimensions which are impossible to produce by other methods.

F. Larger parts: Part sizes are limited only by the size of the isostatic pressure chamber.

G. Low equipment cost.

H. Low die cost: For short runs, the die (mold) cost is low.

I. Small space requirements.

Flexibility: For production situations requiring rather frequent die changes, isostatic pressing offers a definite advantage. Dissimilar shapes can be pressed at the same time simply by placing different molds into the isostatic chamber together.

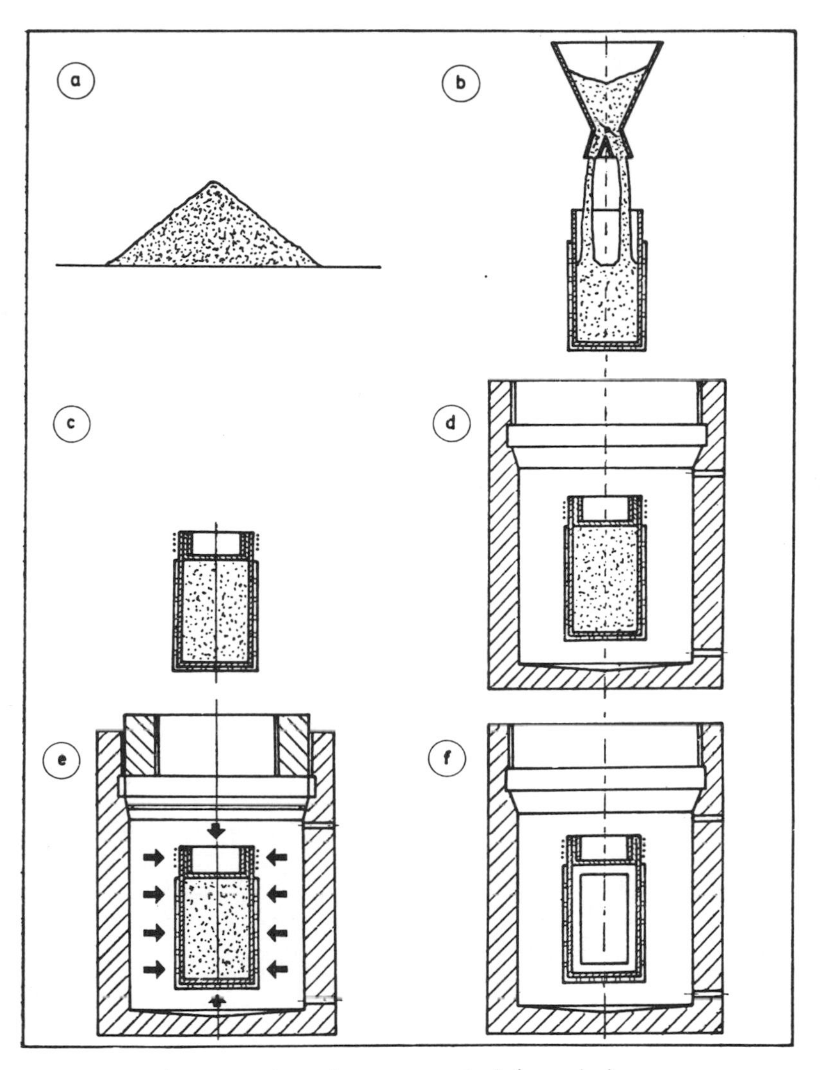

a) material to be compacted (powder)
b) fill-up of flexible form
c) closed and sealed form
d) form in pressure medium in pressure vessel
e) pressurising
f) resulting compact, after decompression

FIGURE 23-1
Isostatic pressing-schematic.

K. Hard to press materials: Some materials are impossible to press and obtain satisfactory results except through the use of isostatics. Many powders can be pressed dry without water, lubricants, or binders.

23.12 Isostatic Pressing Equipment

A. Equipment considerations: An important consideration in any isostatic pressing system is the design of the equipment, and particularly the design of the pressure chamber. The pressure chamber must be designed to withstand the severe cyclic loading imposed by rapid production rates and must take into account fatigue failure.

Equally important is the design of the cover or closure. Not only must it reliably withstand high pressure and cycling conditions, but also it must be designed for quick, easy opening and closing. Several types of closure designs are being used currently to meet these requirements, including threaded pin and hydraulic clamp types. These are available both in manual and automatic systems.

Pressure vessels designed and constructed per Section VIII, Division 2, of the ASME Code are available for pressures up to approximately 40,000 psi (276 N/mm^2). This is now a requirement in most states. When vessels are expected to operate in excess of the ASME code, most states require special permits from the state prior to operating the system.

B. Pressure vessel safety: Cyclic fatigue is the most typical form of pressure vessel failure. A fatigue crack, once started, grows at a determined rate when subjected to repeated pressurization cycles. When the size of the crack becomes critical, the vessel fails catastrophically.

The important thing to remember is that this happens after a large number of cycles over a period of time. Periodic in-service inspection of the pressure vessel, using nondestructive techniques, is the single best method for detecting a crack before it becomes critical. The pressure vessel supplier should be able to provide guidance in this area and may provide inspection services.

23.13 Isostatic Pressing Cycle

The function of each component in the CIP system can be understood better by following an example of a complete pressing cycle. The figure 23-2 is a schematic diagram of a typical CIP unit.

A. Insert compact: The mold which has been filled with the material to be pressed is sealed and placed in the partially filled vessel. This causes the liquid level in the vessel to rise.

B. Fill and vent: The upper closure is installed and locked. Any air remaining above the fluid level must be removed. Air is highly compressible and would consume great quantities of energy before the vessel reaches operating pressure. The fill and drain system adds low pressure water to the vessel, and the displaced air is vented through the valve in the top closure. In this manner the vessel is "topped off."

C. Pressurize: After all air has been vented, the high pressure pumping system pressurizes the water to the operating pressure. Water is added during pressurization to equal the amount of volume reduction of the powder being pressed and to compensate for the compressibility of water at these pressures. A glance at figure 23-3 shows the compressibility of water at various pressures.

Pump up times are an important consideration in total system cycle time. This is a function of the pumping system flow rate at a given pressure. The faster pump up times require more expensive pressurization systems.

D. Depressurization: During depressurization, that volume of water which was added because of the compressibility of water is expelled from the vessel. The simplest decompression controls consist of one hand-operated shut off valve and one hand-operated orifice valve. There are also sophisticated proprietary depressurization systems capable of almost any decompression profile to help eliminate compact breakage that may occur by too rapid depressurization.

E. Remove compact and drain system: The green compact and mold are removed from the pressure vessel. The water level drops some, but would overflow if another uncompacted mold was placed in the vessel. The fast fill and drain system is used to further reduce the vessel fluid level by that volume of water which was added to compensate for the volume reduction of the powder during pressing.

23.14 Tooling for Isostatic Pressing

Isostatic pressing uses flexible tooling of neoprene rubber, urethane, polyvinyl chloride or other elastomers. Rubber is widely used, particularly when thin wall tooling is required. It is used also for pressing more complex shapes when more rigid tooling would present problems in extracting the pressed part from the mold, and in applications calling for disposable tooling. Where more rigid tooling is required, polyvinyl chloride and urethane should be specified.

A. Thin wall tooling: Figure 23-4 shows a thin wall tooling assembly. The flexible mold is made using a form which is dipped into rubber, plastisol, or neoprene. This type of mold is very flexible and is suited for highly compressible material; however, it must be supported for handling and proper shape control. This type tooling is very inexpensive.

B. Rigid tooling: Figure 23-5 shows a thick wall rigid tooling assembly. The rigid tooling is usually made by casting urethane into a two part form. The principal advantages of this type tooling are good shape definition, mold

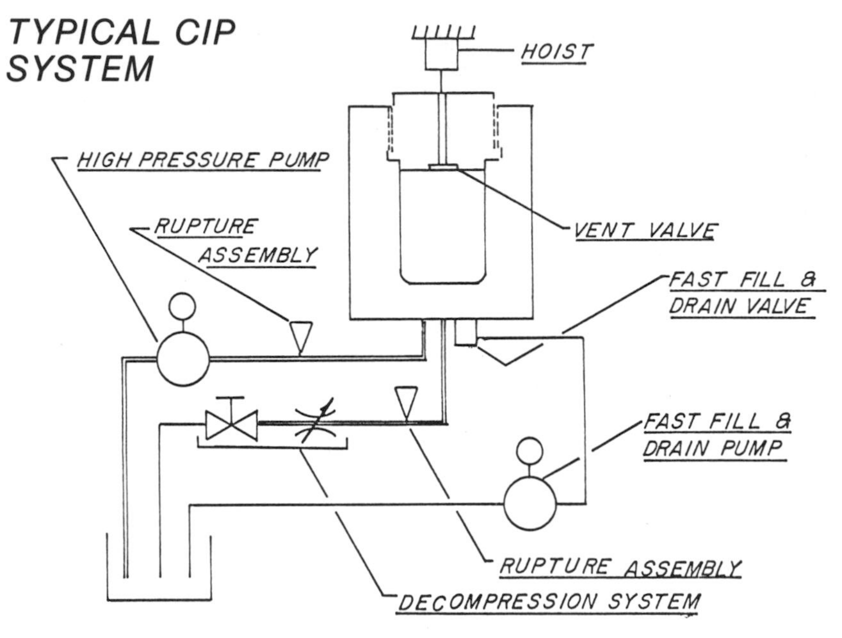

FIGURE 23-2
CIP system-diagram.

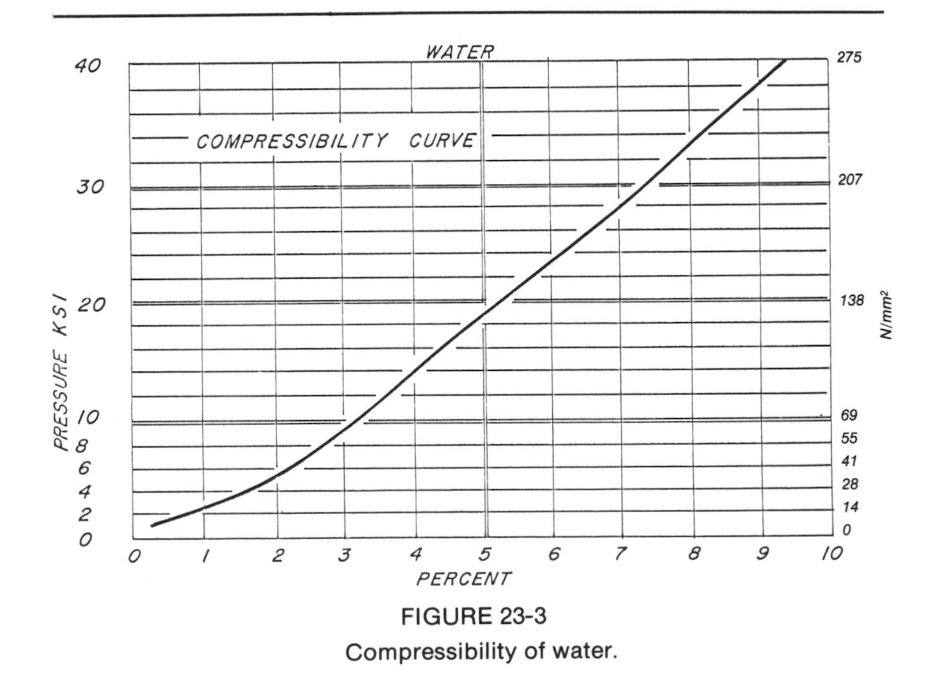

FIGURE 23-3
Compressibility of water.

durability, chemical resistance to commonly used pressing fluids, and the mold usually does not require a support canister.

The drawbacks of the rigid type mold are the expense, highly compressible materials cannot be pressed in this mold, reentrant shapes cannot be pressed unless the part can be removed from the relaxed mold, and adhesions of certain powders to the mold wall cause breakages of the part upon decompression.

Regardless of the type mold selected, compact breakage sometimes occurs. This can be caused by one or several of the following:

- Entrapped air in compact. Vibration during filling, or removal of air by vacuum may be necessary prior to pressing.
- Rapid depressurization. Decompression must be slowed.
- Adhesion of a heavy mold to the compact.
- Poor green strength. Binders may be necessary in some cases.

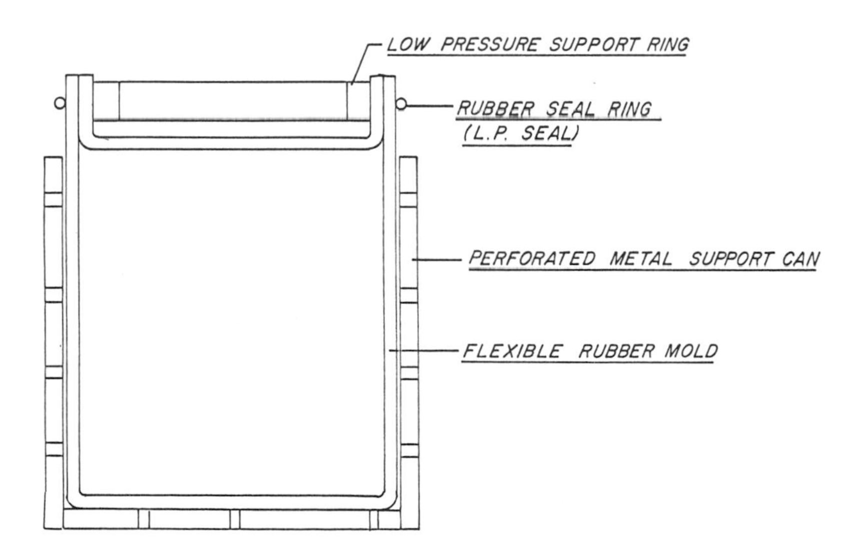

FIGURE 23-4
Thin wall CIP tooling.

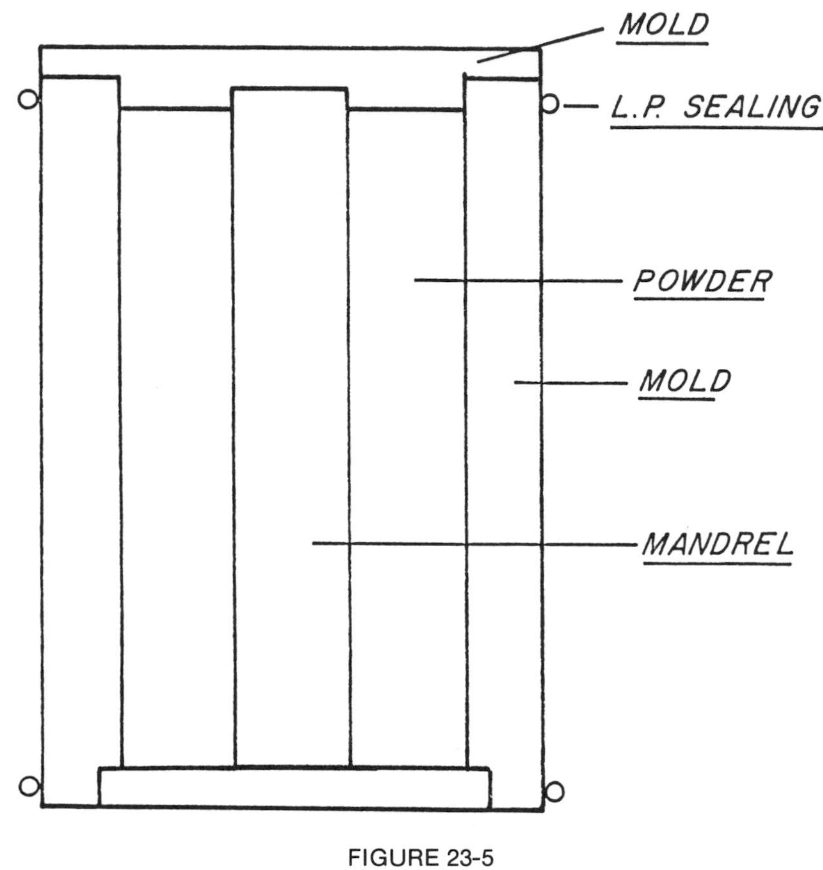

FIGURE 23-5
Thick wall CIP tooling.

23.15 Technical Considerations

A. Pressure/density relationship: Compared with cold pressing, isostatic compaction applies pressure uniformly over the entire surface of the mold. Furthermore, if necessary, air can be evacuated from the loose powder before compaction. Consequently, isostatic compaction provides increased and more uniform density at a given compaction pressure, and relative freedom from compact defects when applied to brittle or to fine powders. Because of the uniform compaction pressure the cross section to height ratio of the part is not a limiting factor as it is with cold pressing. Another characteristic of the process is the elimination of die wall lubricants. Their absence permits higher pressed densities and eliminates problems associated with lubricants removal prior to or durng final sintering.

Generally, there is a distinct advantage in using isostatic pressing except for aluminum and iron compacted to high densities. At high densities both die and isostatic compaction produce similar green densities with iron and aluminum powders. For materials such as aluminum that have constant shear stress, the radial pressure becomes approximately equal to the axial pressure, i.e. approaches an isostatic pressure distribution. However, for materials like copper where yield stress is a function of the normal stress on the shear plane, the radial pressure remains less than the axial pressure. Although the pressure distribution within a cold pressed compact may become isostatic, presumably the pressure vs. density relationship should be identical with that of isostatic compacting only if the density distribution is equally uniform.

Because with isostatic compaction the effects of die-wall friction are absent, much more uniform densities are obtained. This is in contrast to cold pressing where it is well established that die-wall friction exerts a major influence on density distribution in the absence of radial powder flow.

23.20 Hot Isostatic Pressing - HIP

In principle, powder is placed in a sealed mold then subjected to inert argon or helium gaseous isostatic pressure at an elevated temperature between 950F (510C) to 3600F (1980C) with pressures ranging from vacuum to 15,000 psi (100 N/mm^2) depending upon the product being processed. Temperatures are usually selected to allow solid state diffusion to take place but to avoid any liquid phase. In some complex alloys or mixtures a partial liquid phase is unavoidable, particularly in binder materials. Close temperature and pressure control are essential to obtain consistently reliable results.

Rubber or other conventional elastomers cannot withstand the imposed heat so metal or glass are used as tooling. This automatically restricts shape and configuration to relatively simple forms. Thus, the process is normally carried out in the free mold manner. Some experimental work has been done

Table 23-I
Typical Pressures Required for Isostatic Pressing Powders

	Ksi	N/mm^2
Aluminum	8-20	55-138
Iron	45-60	311-414
Stainless steel	45-60	311-414
Copper	20-40	138-276
Lead	20-30	138-207
Tungsten carbide	20-30	138-207

These pressures are approximate since every application presents its own unique requirements of density, configuration and size.

to produce complex configurations using a quasi-isostatic tooling system wherein a high temperature ceramic such as alumina is used as the rigid member.

23.21 HIP Applications

The application of the HIP process has proven that many different materials achieve superior quality characteristics. A wide variety of materials must be "HIPed" because there are no other comparable cost-effective forming processes available today.

A. Densification of cemented carbides

HIP is a particularly effective process for removing internal flaws from sintered carbides. The carbide powder is cold pressed, dewaxed, sintered, then subjected to HIP to eliminate residual porosity. It is widely accepted that for critical applications where increased strength and decreased porosity are required, the HIP process greatly improves all carbide materials. Typical carbide HIP cycles are performed at pressures to 15,000 psi (100 N/mm^2) and temperatures to 2730F (1500C).

B. Consolidation of metal powder

Processing of metal powder shapes, mainly superalloy, tool steel, and titanium, by HIP consolidation has been used extensively since 1970. Three major uses for HIP consolidation of powder are: billets followed by forging/extrusion, semi-finished shapes followed by forging, and near-net shape.

The container material usually is fabricated from stainless or carbon steel. After the HIP process, the container is removed by chemical etching or machining. Containers must be fabricated to high quality standards to avoid an aborted cycle, loss of powder or furnace damage. Ceramic and glass containers also are used for some applications.

Powder consolidation requires temperatures to 2300F (1260C) and pressures to 30,000 psi (200 N/mm^2), although for many applications 15,000 psi (100 N/mm^2) is adequate. Thermocouples must be attached to the workpiece to ensure accurate temperature monitoring on every location. Since the powder is containerized, gas purity is important only to safeguard the furnace from premature failure.

Hot isostatic pressing produces a form close to its final shape, thereby reducing scrap. Uniform powder dispersion throughout the workpiece cross section and zero voids ensures that the HIPed part will have high mechanical integrity. These forms can be used reliably for highly stressed applications.

C. Sinter HIP

Conventional sintering produces parts containing some micro and macro porosity which are detrimental to strength or to surface integrity when polished. Sinter HIPing is used directly for vacuum or atmospheric sintering of parts with the introduction of gas at the end of the sinter cycle.

This process is particularly useful for tungsten carbide, superalloy, and ceramic products where the HIPing process is used extensively. The sinter HIP eliminates the need for reheating the product to liquidous temperature a second time, which risks abnormal grain growth.

The major benefits of combining the two processes are to eliminate one major piece of equipment, reduce materials handling, provide greater control of the process, and save energy and labor costs.

23.22 Equipment

A HIP system usually consists of five major components: pressure vessel, internal furnace, gas handling, electrical, and auxillary systems (fig. 23-6).

HIP systems range in size from 1 to 60 in (25 to 1524 mm) diameter. The smaller units usually are used for research. It is common practice to design one unit universally for research processes such as: densification of ceramics (2000C), HIP of cemented carbide (1500C), consolidation of superalloy powder (1250C) and carbon impregnation (1000C). This can be accomplished with one basic system but with various plug-in furnaces and a versatile control system.

Larger size production units are usually designed for handling a specific process. Standard production equipment for most of the processes listed previously are available.

A. Pressure vessel

Most HIP vessels have threaded closures. They consist of vessel body with threaded top and bottom covers.

The pressure vessel in a HIP system contains the high temperature furnace and retains the high pressure gas. Utility connections to the furnace extend through the bottom cover and require a reliable pressure seal as well as electrical isolation from the vessel. The sealing of the gas in the vessel is done with elastomer O rings since the vessel temperature is kept below 480F (250C).

The design of any pressure vessel must meet two essential requirements. They are minimum dimensions based on allowable stress and fatigue life. The ASME code Section VIII, Division 2 requires a minimum 3:1 safety factor of stress versus tensile strength. Figure 23-7 shows the design fatigue curve for low alloy steel. The curve is based on actual strain fatigue testing applying a

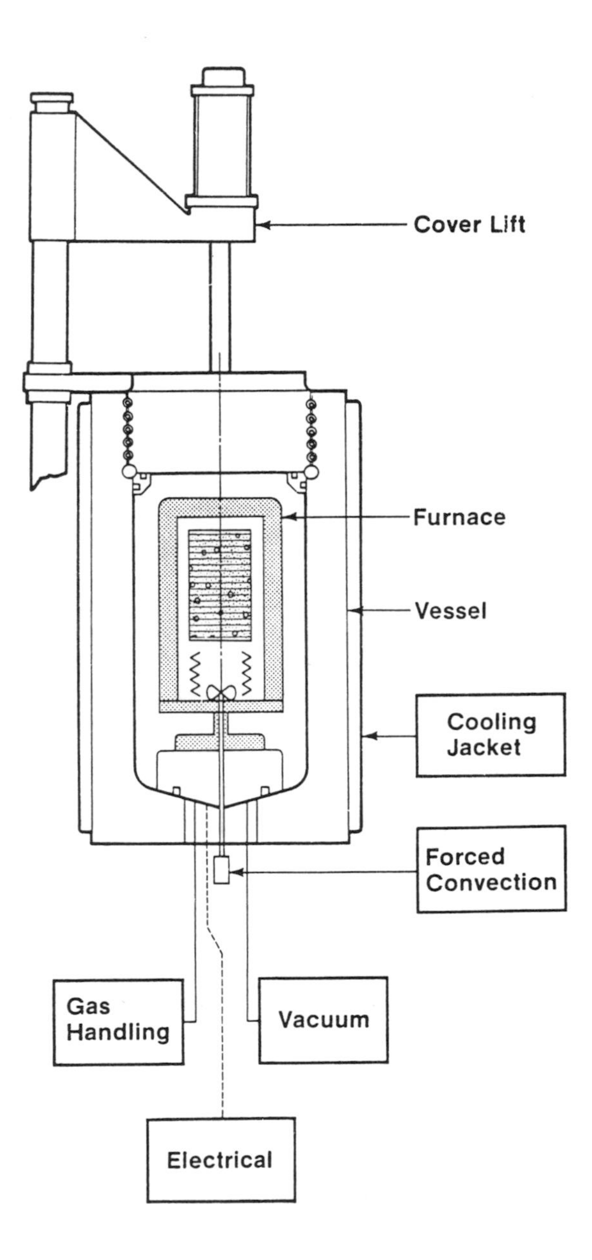

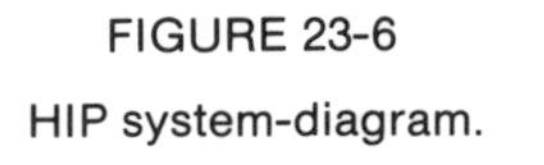

FIGURE 23-6

HIP system-diagram.

safety factor of 20:1 on cycles, or 2:1 on stress, whichever is more conservative.

The quality assurance of a vessel after installation is very important, regardless of the type, size, design, and stress analysis. Once the vessel has been installed it is recommended to establish an ongoing quality assurance program as follows:

1. Annual visual and dimensional vessel inspection with liquid penetrant examination of maximum stressed areas to be sure the vessel is free of surface defects.
2. Ultrasonic examination of the vessel to detect subsurface cracks after every 25% of the design cycle life or every 5 years, whichever comes first.

B. Furnace

The HIP furnace contained within the pressure vessel provides the heat required either from direct radiation or from gas convection. Within the furnace are electrical resistance heating elements and a space for placing the workpiece. The pressure vessel is designed as a "cold wall vessel" and requires protection from the high temperature. A thermal barrier provides this protection and prevents hot gas penetration to the inside vessel wall. Furnaces can be constructed to "plug-in" with the workload in place. Direct thermocouple attachment takes place outside the vessel.

1. Radiation furnaces are typically multi-level, multi-zone styles with the heating element surrounding the workpiece. There are two types of radiation furnaces used; the cold load system where the element and

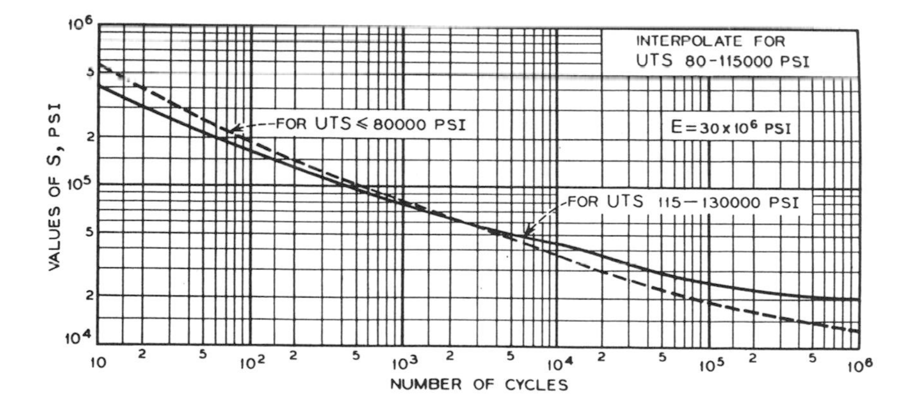

FIGURE 23-7
ASME fatigue curve-low alloy steel.

workpiece start at room temperature and are heated together, or the hot load system where the workpiece is preheated outside the vessel then loaded into the hot furnace cavity.

2. Natural convection furnaces, available since 1976, are common for many sizes of HIP units (fig. 23-8). They work by heating the dense gas in the furnace element area and convecting it to the workpiece above by the buoyancy of the hot gas molecules. In this type furnace a convection liner creates a path for the gas flow as its energy dissipates to the workpiece and as more hot gas molecules move upward from the element. This gas circulation continues until the temperature is equalized throughout the work area.
3. Forced convection furnaces are also of single level construction but have a fan to circulate the gas. Heat transfer to the workpiece is a function of the full coefficient of heat transfer of the gas. By increasing the gas velocity the final coefficient is increased to provide high heating/ cooling rates. The limitations for a system are now only a matter of furnace power for heating and the pressure vessel heat flux capability during cooling.

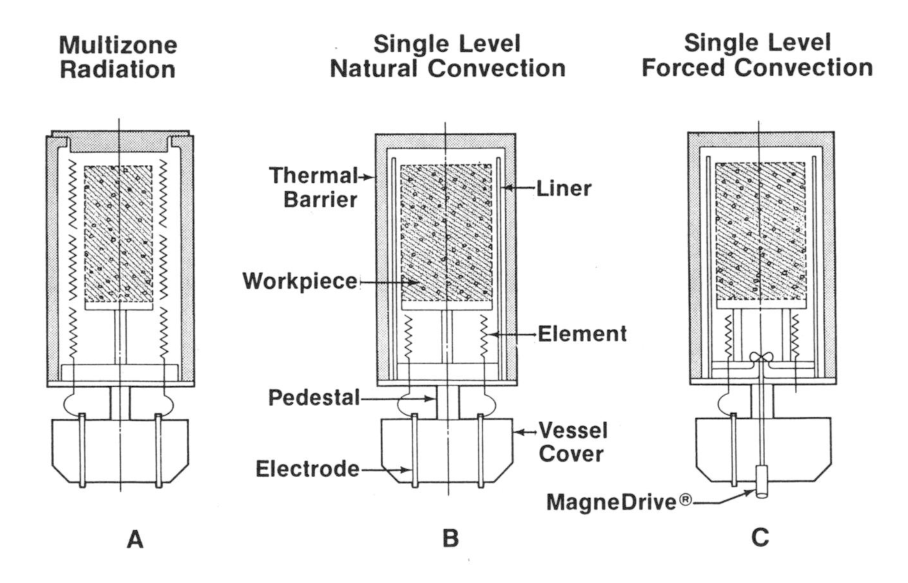

FIGURE 23-8
HIP furnace styles.

There are several advantages to using convection furnaces. The workpiece is not exposed to direct radiation from heater elements. There is a larger work cavity available for a given vessel diameter. Heating elements are not susceptible to damage by the load/unload process, and construction is much simpler than that for multi-level furnaces.

Many different furnace element materials are available for HIP furnaces (Table 23-II). Each element material has characteristics that affect its capability in HIP applications. The three most common element materials are graphite, molybdenum, and nickel/chrome.

C. Gas handling

HIP processing requires an inert gas to apply an equal force (iso) to the part for densification. Most systems use Argon as the pressurizing medium. Helium is still used in a few installations.

HIP systems often require pressures to 14,500 psi (100 N/mm^2) and even up to 29,000 psi (200 N/mm^2) depending on the material being processed. Gas pressures can be achieved with a compressor or by thermal expansion.

1. Diaphragm compressors are most common for pressures up to 14,500 psi (100 N/mm^2). They are motor driven and have a reciprocating piston that develops oil pressure to deflect the metal diaphragm. Triple diaphragms with leak detection monitors ensure that oil cannot enter into the gas stream. The diaphragm life is from 200 to 500 hours depending on gas cleanliness and service conditions.

A combination of two separate compressors has been used for pressurization and reclaim. The primary compressor is rated 3600 psi (25 N/mm^2) and is used to boost the secondary compressor rated 14,500 psi (100 N/mm^2) for increased capacity (fig. 23-9). The primary compressor can also be used to reclaim 90% of the gas from the vessel.

2. Non-lubricated piston compressors are electro/hydraulically driven single or dual stage, and rated to 29,000 psi (200 N/mm^2). The drive fluid is segregated from the process gas to avoid contamination. Seal life is about 500 to 1000 hours, depending on the service conditions.

Liquid pumps are attractive for larger capacity requirements on production systems up to 1000 N/mm^2/hr. for 20,000 psi (138 N/mm^2) pressures. These compressors are motor driven, reciprocating piston pumps, and compress the fluid in the liquid state which is then vaporized before it enters the vessel. The seal life is about 1000 hours depending on the service conditions.

One drawback of liquid pumps is their inability to recover the gas. Gas recondensing has not been demonstrated to be cost effective to date.

Gas purity is very important when processing parts such as castings which are susceptible to oxygen, hydrogen, carbon monoxide, carbon dioxide,

TABLE 23-II
HIP Heating Elements

Heating Element Material	Max. Temp. Rating*F (C)	Advantages	Disadvantages
1. Ni/Cr - Alloy	2150 (1175)	1,7	2,3,4,5,6,7,10
2. Cr/Al/Fe - Alloy (Kanthal/Hoskins)	2345 (1285)	1,7	2,3,4,5,6,7,10
3. Molybdenum Disilicide	3100 (1700)	Not recommended for HIP	
4. Silicon Carbide	3000 (1650)	1**,4,5,8	2,3,7,9
5. Molybdenum	3100 (1700)	2,6	1,3,4,6,7,8,9,10,11
6. Tungsten	3450 (1900)	2,6	1,3,4,6,7,8,9,10,11
7. Graphite	5450 (3000)	2,3,4,5,6,8	1,11
8. Tantalum	4000 (2200)	2,6	1,3,4,6,7,8 9,10,11

Explanation of Advantages/Disadvantages

Advantages

1. Operates in most atmospheres like: inert, air, O_2, etc.
2. Watt density greater than 10 W/cm^2.
3. Rapid thermal cycling.
4. Elements are self supporting.
5. No grain growth with thermal cycling.
6. High strength and dimensional stability at temperature.
7. Low initial cost.
8. Low maintenance cost.

Disadvantages

1. Operates in vacuum or inert atmosphere only.
2. Watt density lower than 10 W/cm^2.
3. Prone to thermal shock.
4. Elements require special ceramic support.
5. Large grain growth with thermal cycling.
6. Embrittlement with thermal cycling.
7. Metallurgical change develops hot spots.
8. Resistance change $>$ 5:1 due to temperature.
9. High initial cost.
10. High maintenance cost.
11. Susceptible to contaminants like O_2, CO_2, CO, H_2O

*Maximum temperature rating depends on element design and electrical isolation support.
**No data for high pressure air or O_2.

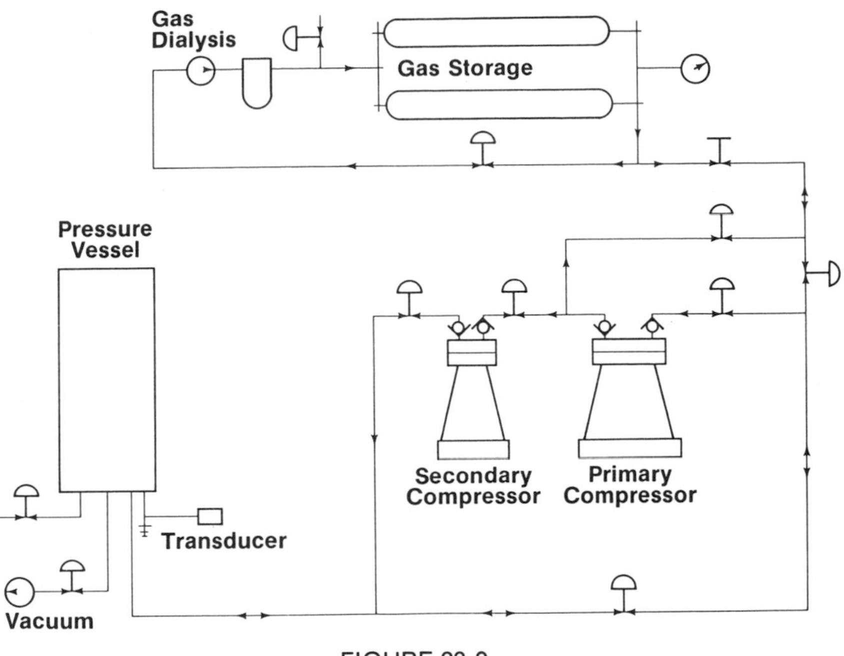

FIGURE 23-9
Gas flow-schematic.

water vapor, and hydrocarbons. Gas with 99.995% purity (50 PPM total impurities) is acceptable to prevent premature furnace failure, but can be harmful to superalloys or titanium which require less than 5 PPM total impurities. The gas can be purified with gas dialysis equipment or by gettering in the HIP furnace. On-line gas analysis is a must when processing surface-exposed parts.

D. Controls

The control system for a HIP unit is the subsystem that links the vessel/furnace, gas handling, and auxiliaries into a functioning research or production tool. Computer control ensures repeatibility of these parameters to maintain consistency of operation.

HIP equipment requirements dictate that a versatile control system is necessary. Research and laboratory scale equipment is available with furnace styles that vary widely in construction and operation. These interchangeable furnaces have different elements and thermocouples operating over a wide range of temperatures. The mini and micro computer control allows for the changes by programming and software. This provides the scientist an opportunity to operate throughout the whole range of cycle parameters with accuracy and repeatibility.

The heart of the control system is the control panel. Operation is performed by a sequence of events initiated at the control panel by some form of logic. The common logic types available are relay, programmable controller (Level 1), mini and micro computer (Level 2), and total computer supervision (Level 3).

1. Relay and programmable controller logic are similar to controls used in many industrial environments. These units typically require continual operator interface to provide the judgment and initiation to proceed with each step of the HIP cycle.

2. The mini and micro computer controls allow complete automatic control of HIP cycles. The cost of a single microprocessor chip has steadily declined until mini and micro computers are cost competitive with relay logic and programmable controllers.

3. Total computer supervision is provided by a highly sophisticated system that operates HIP production equipment without an operator interface. This system includes load/unload systems with machine robots for materials handling. This system handles every phase of HIP production from the raw material to the finished product.

E. Auxiliary systems

A HIP system is supported by a number of important subsystems classified as auxiliaries. These include a cooling system and vacuum system, material handling with workpiece fixtures and facilities subsystems including exhaust fans, oxygen monitoring equipment, and cranes.

1. The cooling system keeps the pressure vessel temperature below its design limits. Closed loop cooling is used with treated water to prevent corrosion of low alloy steel pressure vessels. Since cooling is essential for energy removal from the vessel, the system is provided with temperature and flow sensors.

2. The vacuum system provides a means of removing the atmospheric contaminants from the furnace/vessel. Commonly a mechanical pump with a Roots type blower is used for handling high flow rates at low pressure. Isolation is provided by a high capacity/high pressure vacuum valve. System interlocks ensure safety and prevent exposure of the vacuum components to high pressure.

3. Material handling can be integrated into the furnace design. The furnace can be loaded outside the vessel, the thermal barrier placed over the workpiece and locked in place. The furnace and workpiece are then lowered into the vessel as a module. Alternatively, the thermal barrier can be removed

and the workpiece lowered into the vessel on a fixture. Either of these methods uses a crane for handling the furnace load.

4. Workpiece fixturing is determined by the material being processed as well as by the operating temperature. Workpiece fixtures for various HIPed materials vary but commonly used materials for superalloy billets and castings are materials that maintain strength at high temperature. Carbon steel has a relatively high melting temperature, and is used in some applications. Materials such as nickel/chrome alloys with high creep strength are not prone to severe phase changes. They can be used also for workpiece fixtures. For processing tungsten carbide, where high temperatures are required, a purified graphite fixture is used. This 4000F (2200C) HIP application requires a combination ceramic and graphite fixture.

5. Oxygen monitoring equipment is used with inert gases which are heavier than air and do not support life. Strict safety precautions must be followed. It is possible to be asphyxiated in the confined space of a pit or pressure vessel where the air has been displaced by the inert gas. Oxygen monitoring and exhaust equipment become essential for the safety of personnel. The monitoring equipment can be set up to automatically start exhaust fans and give warnings of an oxygen deficiency.

23.23 Typical Cycle

The powder to be compacted is placed inside a container or mold, usually made from mild steel. A top fitted with a suitable evacuation connection is then welded in place.

Evacuation and outgassing of the powder is accomplished by connecting a vacuum pump to an evacuation connection extended to reach outside the preheating furnace. The container is pumped down until a level of at least 1,000 microns (1 Torr) is reached and the work is then placed inside a preheated furnace and brought up to 800 to 1200F (425 to 650C). Once temperature is reached, vacuum pumping is continued until the 1,000 micron (1 Torr) level is reached inside the container. The work is then removed from the furnace, still under vacuum, the connection tube is crimped and welded shut close to the can. The excess length trimmed off to permit efficient loading into the hot isostatic press. These two steps can be done independently of the HIP unit so loads can be prepared in advance and stockpiled to keep the HIP unit used efficiently.

One or more prepared cans of material are loaded into the work space of the HIP unit. The number and arrangement of the cans is limited only by the necessity to ensure that the desired temperature be maintained uniformly and that adequate mechanical support is provided. Conventional alumina kiln furniture is used for some parts. Crucibles and/or boats can handle small parts. Various grades of alumina powders and balls can also provide the required mechanical support. It is common practice to bench pack a large mild steel cylinder, which is sized to fill the work space of the system, with as many parts as feasible so that only a single element need be inserted and removed from the chamber. In small laboratory size equipment, loading is done directly into the HIP unit, but the method described previously is used in all larger systems.

The HIP unit is closed and locked, evacuated to minimize contamination of the pressurizing gas, then the pressure and temperature are usually raised simultaneously. When glass or other brittle canning materials are used minimum temperature levels are established first to ensure that the canning material is adequately ductile before pressure is applied. In other cases, the load is pressurized cold to an intermediate level then the pressure is raised to the designed working level by increasing the temperature.

After reaching the desired pressure and temperature levels, the system is kept at a "hold" or soak condition to ensure that the center of the load has reached the set conditions. No generalizations can be made covering the time required since that is a function of the load characteristics, shape, mass, composition and configuration of loading. Thermocouples placed in the load indicate completion of the minimum soak time. Additional time might be specified to achieve particular metallurgical properties. These are separately evalutated on the basis of diffusion rates and similar factors in a given application.

Finally, pressure and temperature are allowed to drop. Normally the furnace is merely shut off completely and the temperature is allowed to decrease at its maximum rate. Pressure is usually allowed to drop in thermodynamic ratio to the temperature drop since bleeding off gas to reduce the pressure will slow down the cooling rate. Some materials require a slower, controlled cooling which can, of course, be achieved in most HIP systems. Cycles may be timed so that the cool down phase takes place at night. Next morning the system is ready for venting to zero pressure and reloading. This is a normal production cycle condition. There are some operations which can achieve two cycles per day, while others, by the nature of the cycle required by the product might run up to two or three days.

PART

THREE

SINTERING

30.00 SINTERING

31.00 SINTERING FURNACES

A sintering furnace requires a gas-tight furnace shell or a gas-tight muffle to maintain a reducing furnace atmosphere, a heated burn-off or purge chamber to expel air and lubricant vapors, a controlled rate of pre-heat, a controlled high-heat sintering chamber, and a water-jacketed cooling chamber, all with a protective atmosphere to prevent oxidation while the compacts are in the furnace. Figure 31-1 shows a typical temperature profile in a continuous sintering furnace.

31.10 Furnace Types

31.11 Muffle vs. Gas-Tight Shell Construction

Pilot plant and laboratory sintering is carried out in small muffle furnaces. Generally, muffle construction is not used for large production sintering furnaces, unless it is necessary for close control of the furnace atmosphere purity. Large muffles are not only expensive initially, but are also expensive to maintain. Heating through a muffle is less efficient than having the electric heating elements exposed to the furnace atmosphere and furnace chamber. The general rule for large production furnaces is to omit the muffle when using exothermic or endothermic atmospheres. If the furnace is gas fired, then a full muffle must be used to prevent oxidization of the compacts.

Full-muffle construction or a high purity refractory lining is necessary when using hydrogen or dissociated ammonia atmospheres and when dew points of less than -20F (-29C) are required. The successful sintering of stainless steel requires the dew point of the hydrogen to be less than -20F (-29C) at 2100 to 2200F (1150 to 1204C). A full-muffle furnace must be used for this application because hydrogen will react with the silica in the refractories at these temperatures to produce water vapor and raise the dew point.

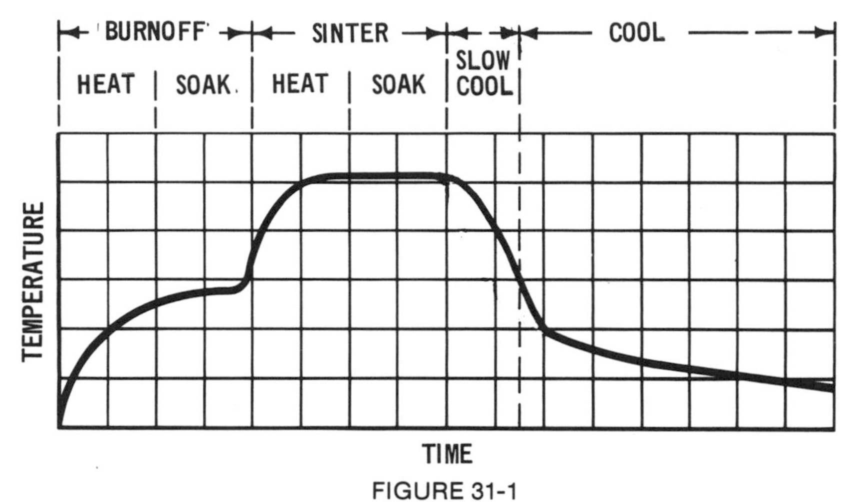

FIGURE 31-1
Typical continuous furnace temperature profile.

An exception to this is the application of very high purity aluminum-oxide refractories used in molybdenum element furnaces, to be described later in more detail. These special refractories are extremely expensive, and the furnace must be operated continuously to obtain and maintain a low dew point atmosphere.

Another advantage of a full-muffle furnace when using expensive hydrogen or dissociated amonia, atmosphere, is that the flow of atmosphere purging gas can be reduced considerably. Full-muffle furnaces can be purged rapidly because there is no entrapped air to be removed from the brickwork. One method of purging a full-muffle furnace is to purge it with dry nitrogen or some other inert gas, so that the hydrogen will not burn to form water vapor when introduced into the muffle. Nitrogen also prevents the muffle from oxidizing and having the oxides converted to water vapor to contaminate the hydrogen atmosphere when it is introduced into the hot furnace muffle.

Another method of purging full-muffle furnaces is to bring the furnace up to a temperature of 1400F (760C) before hydrogen or dissociated ammonia are admitted. The oxides formed in the muffle due to lack of atmosphere gases at these temperatures and the moisture content first developed by the atmosphere gases, can be eliminated during the time that the furnace is being heated to the normal operating temperature.

On small full-muffle furnaces, the maintenance is not too great even at temperatures as high as 2100F (1150C) and, therefore, small furnaces are made with a muffle to expedite purging for intermittent use, and to be universally adapted for use with hydrogen, exothermic, or endothermic atmosphere.

31.12 Mesh-Belt Conveyor Furnace

The most commonly used conveyor for continuous production of small, light parts is a mesh-belt. As shown in figure 31-2 the furnace consists of a charge table for loading the parts on the belt, a purge and burn-off chamber to vaporize the lubricants from the compacts, a high-temperature chamber, an insulated cooling chamber, a water-cooled chamber, and a discharge table. The parts are loaded on a continuously driven alloy mesh belt at the front of the furnace, and are discharged at the rear. A variable speed drive allows flexibility to meet the heating requirements of the work to be treated. The alloy belt limits the furnace operation to about 2100F (1150C). Also, the mesh belt limits the furnace length because the stretch of the belt increases as the total load in the furnace increases. At temperatures of 2000-2100F (1093-1150C) belt loadings are normally 10 lbs/ft^2 (49 kg/m^2) or about double this with a heavy duty belt drive system. At lower temperatures heavier loadings are permissible.

The mesh-belt type of furnace handles low to medium-high production rates, depending upon loading density, heating time, and soaking time. Doors on a mesh-belt furnace usually are left open during operation. In selecting the atmosphere generating equipment, ample capacity must be available.

31.13 Hump-Back Type Furnace

The hump-back mesh-belt conveyor furnace is a variation of the mesh belt conveyor furnace used where high atmosphere purity is required.

A long gas-tight entry inclined purge chamber carries the belt and work from the charge area up to the elevated hot zone (fig. 31-3). Following the hot zone, the cooling section is then inclined downward to a discharge point. Most large hump-back furnaces have a booster drive in the purge section to assist in carrying the work and belt up the incline. This reduces the stress on the mesh belt.

The hump-back furnace is particularly good where light gases such as dissociated ammonia or hydrogen atmospheres are used. These gases tend to be confined in the furnace because of their natural tendency to rise. Because of this characteristic, this type of furnace generally operates with a lower gas consumption than the standard straight-through conveyor furnace. For the same reason it is generally easier to establish a lower dew point and better atmosphere purity in a hump-back furnace.

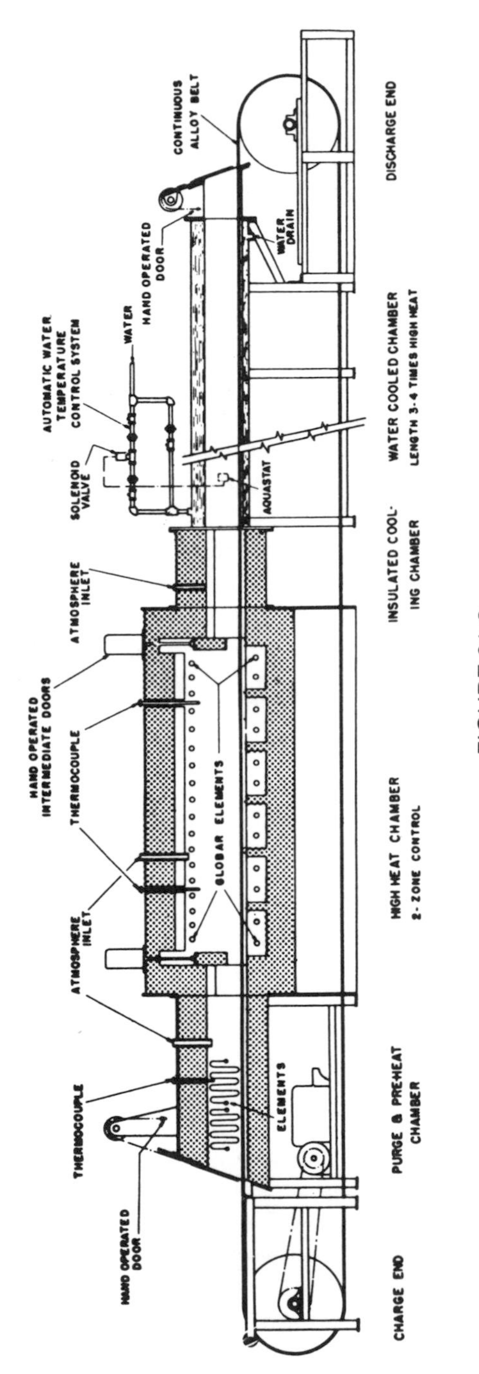

FIGURE 31-2
Longitudinal section of mesh belt furnace.

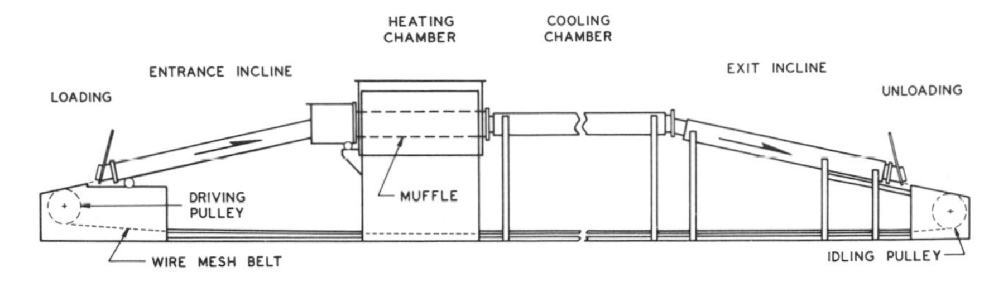

FIGURE 31-3
Longitudinal section of hump-back furnace.

The major applications for this type of furnace are for sintering stainless steel and aluminum, which require a very low dew point.

When a hump-back furnace is used, the burn-off of the compacts often is done in a physically separate furnace to reduce atmosphere contamination. The two furnaces may be linked together with a suitable conveyor to avoid separate handling of the work between the furnaces. An alternate arrangement is to burn off in the entrance inclined chamber.

Like the standard mesh-belt type of furnace, the hump-back is limited to a top temperature of about 2100F (1150C) maximum, and belt stress limits the length of furnace.

31.14 Mesh Belt Maintenance

Nothing is more important to the proper operation of a mesh belt furnace than the belt which carries the parts to be processed. Flatter, thinner spirals in the belt construction and the utilization of new alloys in the construction have all helped advance belt life and furnace technology. The most technologically advanced belt construction will not give proper life if problems encountered in a furnace are not corrected *before* a new belt is installed. Furnace belts do not wear out; they are destroyed by inadequate furnace maintenance.

Edge damage occurs when the belt has wandered or walked to one side and one or both edges become damaged. This is usually indicated by broken edge welds. Pulley adjustment will usually correct this.

Wear patterns on the carrying or the underside of the belt or belt spirals being pulled in a distinct pattern is indicative of something projecting onto one area of the belt and wearing and distorting it.

The most prevalent problem is usually belt camber. Camber can be viewed as the spirals of the belt not running straight across the width of the belt. This will result in premature failure of the belt due to its inability to hinge properly.

Camber can be caused by non-uniform wearing of the belt support surface, non-uniform product loading of the belt or exceeding the load limits of the belt.

Before placing a new belt in a furnace it is most important to check alignment, paying particular attention to the rolls or end drums. The rolls should be checked for roundness, free turning bearings and run out accuracy as they rotate. End rolls should not be crowned or tapered. If the drive roll is lagged, it should be checked for wear; if the drive roll has a press roll, the two rolls should be checked carefully for parallelism. A canted press roll can seriously affect tracking and cause extensive belt damage.

The counter tension roll should be restricted to ensure uniform parallel motion. This roll can also seriously affect tracking, and if not corrected, stretch one edge of the belt and cause permanent belt damage.

All support rolls and snub rolls should be checked for free turning bearings, roundness and straightness (not bent or bowed). These rolls should be adjustable parallel to the direction of the belt travel and are the rolls which should be used to track the belt. End drums or rolls should not be used for tracking.

Rails, skids or other supports over which the belt travels should be checked for smoothness, and squareness with the end rolls. Sharp edges and abrupt corners should be ground smooth or otherwise corrected to prevent any possible scraping or snagging of the belt surface. In some cases the old belt may have failed due to rubbing excessively on these areas; the new belt will fail quickly also if this is not corrected.

All edge guides and other conveyor or furnace parts which could wear the belt edges should be checked. If wear is evident, corrections should be made to be certain the new belt does not drag or wear along its edge.

After checking the conveyor system and making necessary corrections, the belt is ready for installation. While many methods may be used for pulling the belt into the system, it is quite important that the belt be pulled straight and evenly along the center of the conveyor.

Adequate time should be allowed to "break in" a new belt. With the conveyor properly aligned and the counter tension roll in the most contracted and low tension position, start the belt at the slowest speed possible. If the belt tracks radically to one side, it should be stopped immediately, the counter tension should be relaxed and the belt should be positioned back to the center of the conveyor.

As the belt runs, adjust only the support rolls, top and bottom as appropriate, to make the belt track toward the center of the end drums. A metal belt will track perpendicular to the centerline of the support rolls. If a large tracking correction is needed adjust several rolls a small amount rather than one or two rolls a large amount.

Do not adjust end rolls to track the belt and adjust the snub rolls only if the total adjustment of all the support rolls will not give proper tracking.

All belts have some "run out" which should not be confused with tracking variations. "Run out" is the wobble or movement in and out of the belt edge as viewed from any fixed point on the conveyor. Tracking problems occur when a belt tends to run to one side of the conveyor jeopardizing the belt edge. A new belt will not normally track the same as the old belt and it should not be assumed that tracking adjustments are not needed.

Proper "break in" of the new belt is very important. When the belt appears to be tracking well enough for constant running (no danger of tracking off into the side of the system) start increasing the counter tension. This should be done gradually as tracking may change as belt tension increases.

When operating counter tension is applied to the belt, let it operate for several hours tracking in the center of the end drums. Stretching may occur rather rapidly at this point due to the "seating" of the spirals on the cross rods of the belt. "Seating" is the establishment of sufficient contact area between the wires of the belt to sustain the applied tension. Constantly monitor the belt tracking position as "seating" occurs.

When heat is to be applied to the belt raise the temperature slowly, approximately 300 to 350F (150-175C) per hour. Further "seating" will occur as the temperature rises. Constantly monitor tracking as the temperature rises.

For belts that will operate at 1700F (925C) or less, raise the temperature of the belt to the operating temperature plus 50 to 75F (10 to 24C) and let the belt operate long enough to ensure that it is fully stress-relieved. This will normally require four to six hours.

For belts operating above 1700F (925C) stop the temperature rise at 1700F for four to five hours and then increase the temperature to the operating temperature. This procedure will stress relieve the belt properly, preventing premature grain growth of the belt alloy. After the belt has operated for four or more hours at operating conditions make final tracking adjustments.

As the conveyed load is applied to the belt, the belt should be checked frequently during the first few days of operation to detect whether any further tracking adjustments are needed. Further "seating" of the belt may occur during this period and a section of the belting may have to be removed to accommodate the travel distance of the take-up and counter-tension roll.

Over the years different alloys have been developed and used for furnace belts. Most typical are T-314 stainless steel, 35-19 Cb and 80-20 Cb, all typically high in chromium and nickel to help increase oxidation resistance.

Generally T-314 stainless steel is the most economical in pure air atmospheres at temperatures in the 1600F (870C) range; 80-20 and 80-20 Cb alloys are generally more economical in protected atmospheres in the 2100F

(1150C) range; 35-19 and 35-19 Cb alloys are generally a good combination of both.

Belt life can be lengthened by turning the belt over each 1/4 of its expected life and reversing its direction of travel. If the belt is subject to highly carburizing conditions it should be decarburized periodically.

31.15 Roller-Hearth Furnace

In the roller-hearth type continuous sintering furnace, trayloads of parts are conveyed by riding on driven rolls. The charge and discharge doors are automatically opened and closed by air or motor, and are interlocked with the charging and discharging mechanisms. A longitudinal section of a roller-hearth furnace, outlining its components, is shown in figure 31-4. The grade of alloy used in the rolls in the furnace limits its operation to 2100 to 2300 F (1150 to 1260 C). Each roll is driven, and depending on roll spacing, is capable of holding a load substantially greater than an equivalent length of mesh-belt conveyor.

The roller-hearth furnace is built in various widths and in any length to meet medium to high production requirements. The end doors on a roller-hearth furnace are opened only when a tray of work is charged or discharged. This economizes on the amount of atmosphere gas required, and minimizes heat losses.

31.16 Pusher-Type Furnaces

The pusher-type sintering furnace is suited for sintering metal parts that load too heavily per lineal foot for the mesh-belt type, and where the production rate does not warrant the roller-hearth type furnace; or for sintering temperatures up to 3000F (1650C) which is too high for either the mesh-belt or roller-hearth furnaces. Mechanical or hydraulic pusher furnaces are available for high-output capacities. Generally two types of pushing mechanisms are used. One is intermittent pusher type, the other is the continuous stoker pusher type. The intermittent pusher mechanism normally is applied to bronze, brass, and iron P/M parts because of its low cost. The continuous stoker type is applied to furnaces for sintering carbides, stainless steel, and other applications wherein a gradual stepless temperature profile during heating and cooling is important. A typical pusher furnace is shown in figure 31-5 for sintering iron compacts.

31.17 Walking Beam Furnace

The walking-beam sintering furnace is another continuous type that can be used for the production of P/M parts (fig. 31-6). The amount of production

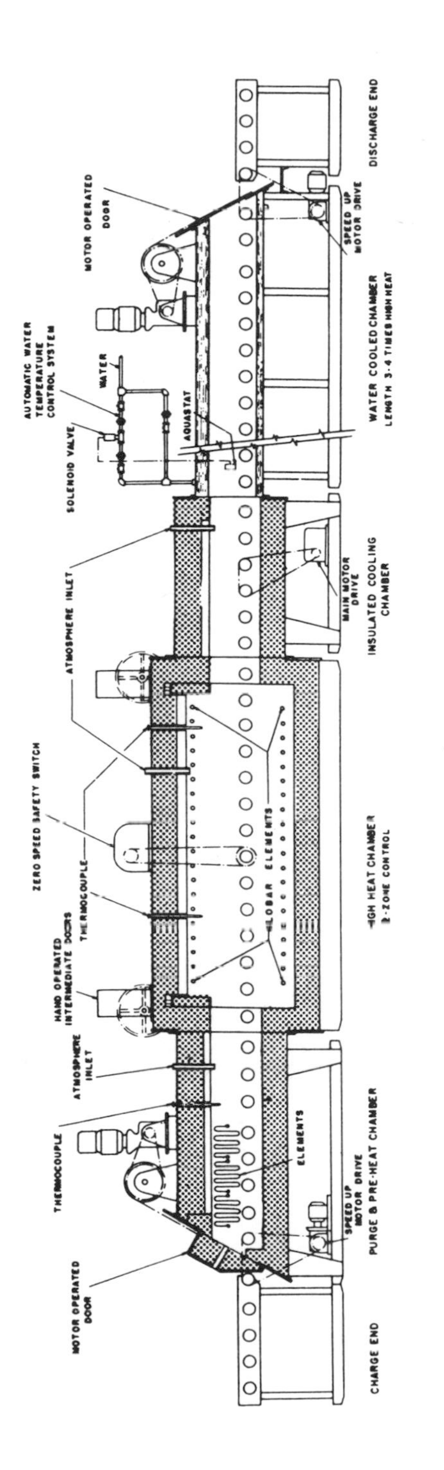

FIGURE 31-4
Longitudinal section of roller-hearth furnace.

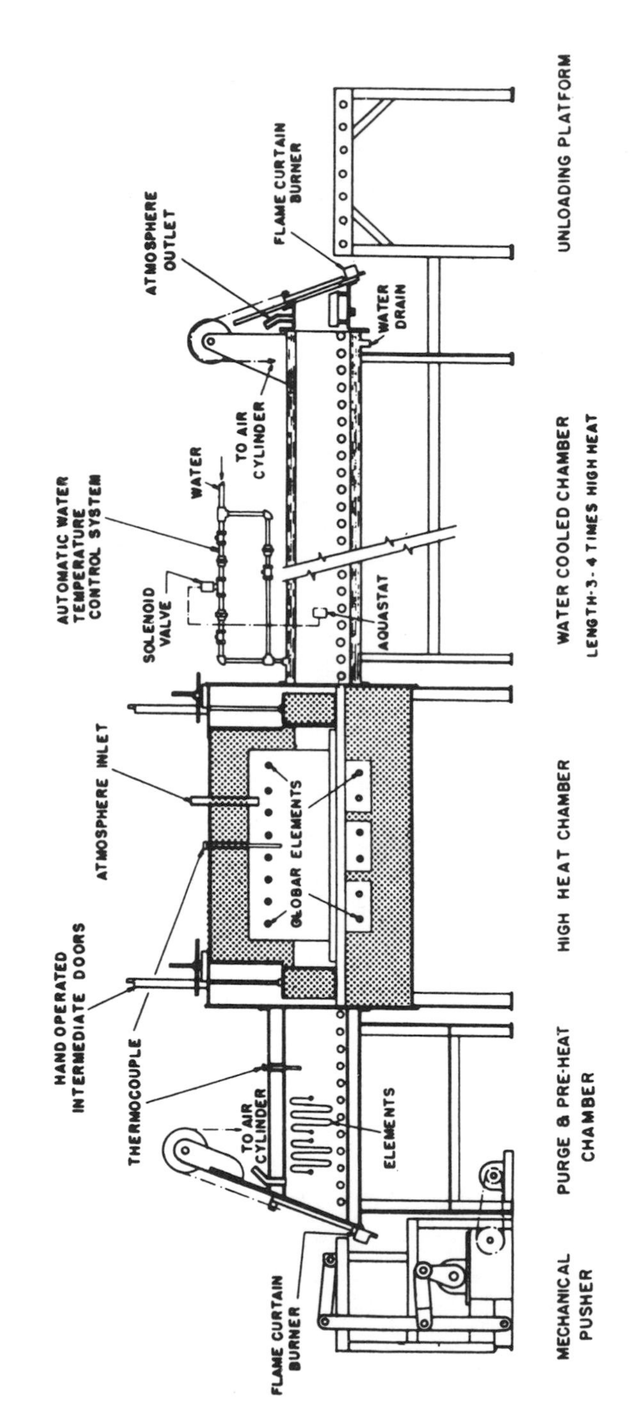

FIGURE 31-5
Longitudinal section of mechanical pusher furnace.

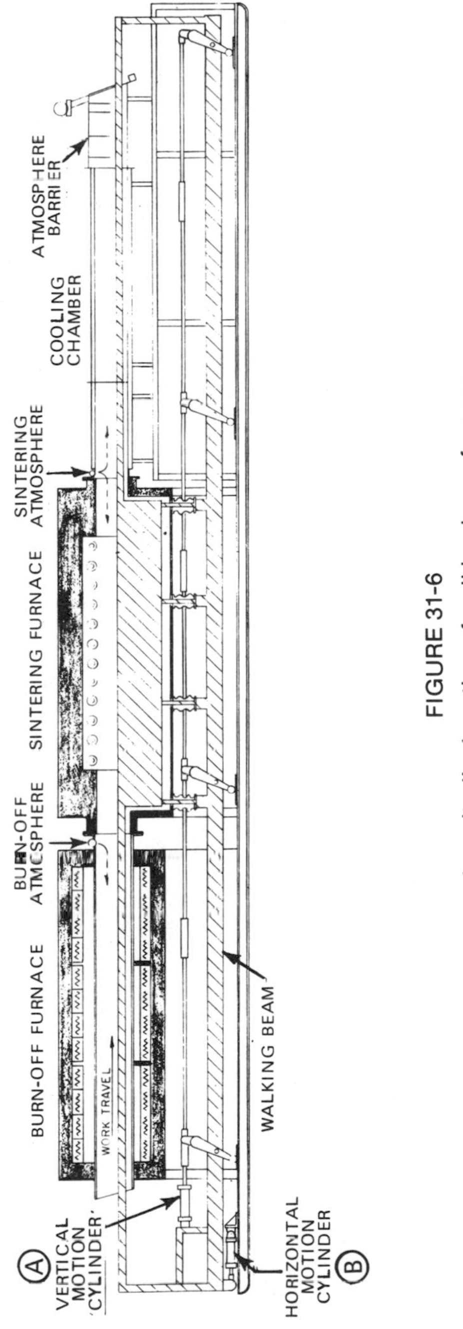

FIGURE 31-6
Longitudinal section of walking beam furnace.

work weight that can be conveyed safely through the furnace is practically unlimited.

The maximum continuous operating temperature of the furnace is limited only by the refractory materials used to line the heating chamber, and by the compatibility of the sintering atmosphere being used with the heating element.

The operating temperatures may be as high as 3300 F (1815 C), depending on the compatibility of the atmosphere and the heating elements used.

The walking-beam is basically a mid-section of the floor throughout the length of the entire furnace, which is raised approximately ½ in. (1.27 cm) above the hearth level, pushed forward a predetermined distance, about 1 in. to 2 in. (2.54 to 5.08 cm), lowered ½ in. (1.27 cm) below the hearth, and pulled back to the starting position (fig. 31-6 and 31-7). The up-and-down motion of the walking-beam is accomplished by means of a hydraulic cylinder (A) operated by an adjustable cycle timer and activated by a four-way valve (double solenoid operated). The cylinder (A) pushes and pulls a lever mechanism which, in turn, raises and lowers the conveyor. A second cylinder (B), which is also activated by the four-way valve (through the same cycle timer as the previous cylinder), provides the forward and reverse motion. The timed combination of the cylinders produces a rectangular motion, conveying the work at the required speed.

Another variation is the dual beam, used in furnaces wider than 18". The basic function of the dual beam is to provide center support for the boat during the lowering and retraction of the beam. Such support shortens the span and allows for lighter boats. Example: a 24" boat resting on 2" shelves would be spanning a distance of 20" with a single beam. The maximum span with a dual beam is only 6¼". The weight of the boat can therefore be reduced by a factor of at least 2.

31.18 Batch Furnace

Where production does not warrant continuous operation, or for experimental work, the pusher mechanism on mechanical pusher furnaces can be eliminated and the work pushed through the furnace one tray at a time, by hand. Muffle or non-muffle batch furnaces with gas-tight interlocks and/or doors for charging and discharging work are also used for critical materials such as stainless steels or Alnico, where the dew point of the hydrogen atmosphere must be kept extremely low. A small hand-pusher muffle batch furnace with low-dew-point hydrogen or dissociated-ammonia atmospheres for sintering yellow brass is shown in figure 31-8.

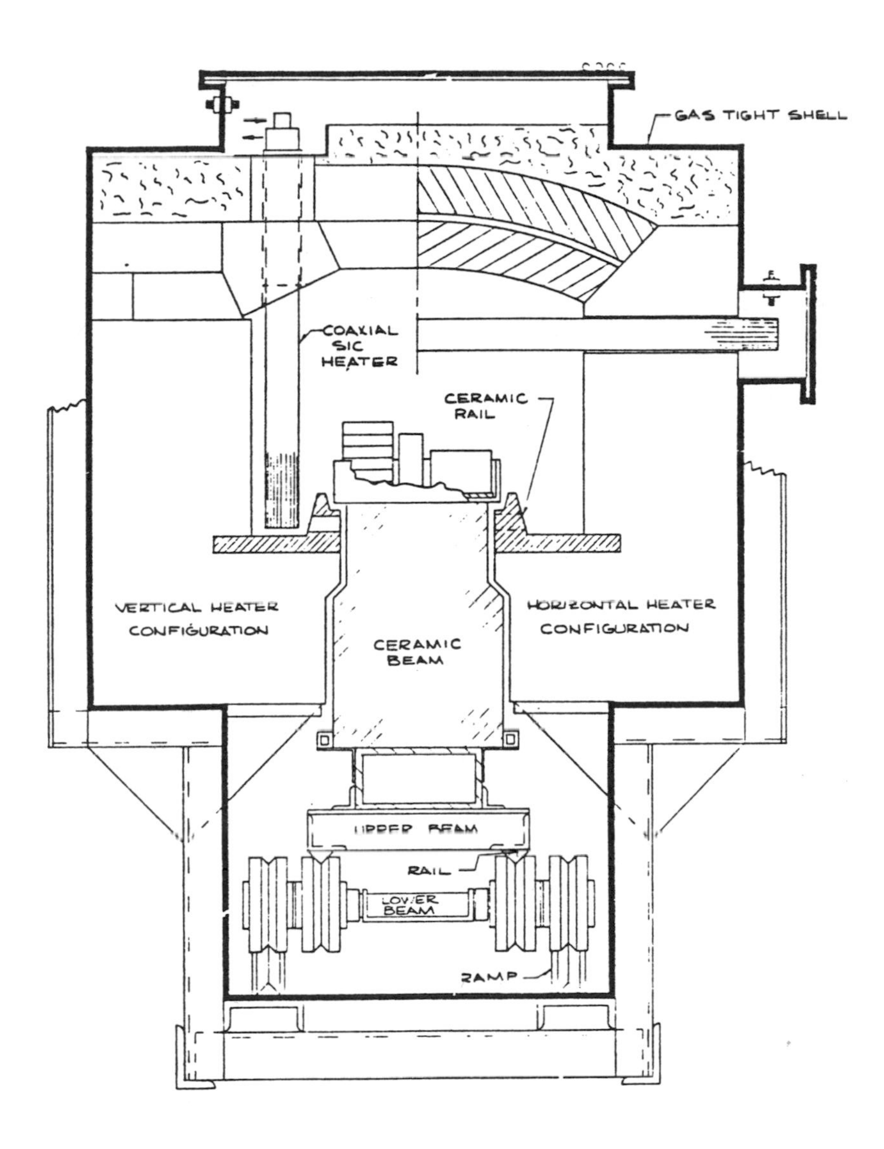

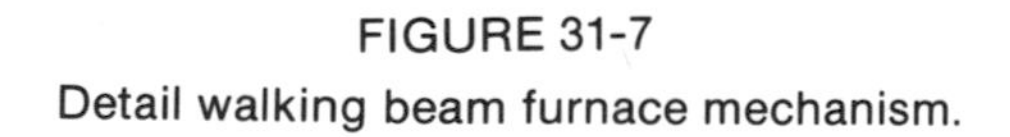

FIGURE 31-7
Detail walking beam furnace mechanism.

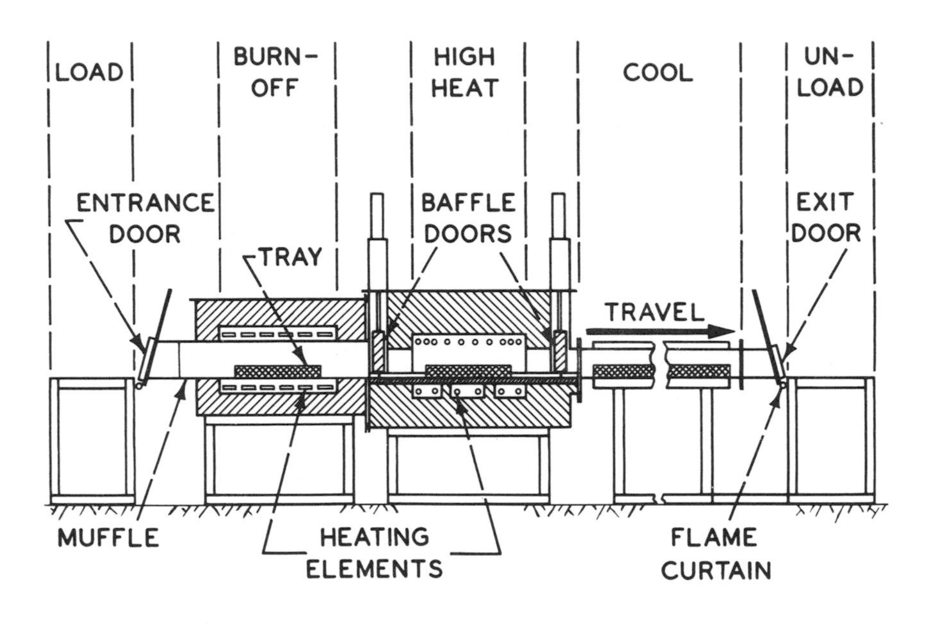

FIGURE 31-8
Longitudinal section-box furnace.

31.19 Bell or Elevator Furnaces

Bell, elevator, or pit type furnaces are used as batch furnaces for sintering. The bell type furnace, figure 31-9, is used widely for long sintering cycles required for P/M friction materials, such as elements for clutches or brakes. Optional pressurizing mechanisms (not shown) are used for applying heavy loads to stacks of work pieces. For example, a pneumatically or hydraulically operated ram extending up through the center of the load; or a pneumatically operated diaphragm on top of the furnace, bearing down on top of the retort and load. Typical equipment consists of one or more stationary load-supporting bases with removable sealed retorts to cover the loads and retain protective atmospheres around them throughout the entire heating and cooling cycles, a portable heating bell and a standby base for it, a travelling hoist, and an optional cooling bell (not shown).

The elevator type furnace, figure 31-10, is useful for sintering heavy or bulky loads, such as pressed-powder billets or stacked trayloads of compacts with work temperatures up to about 2250 F (1230 C). Like the bell type, it is good for applications requiring protective atmospheres of exceptionally high purity, and is flexible for various time-temperature cycles in different types of loads. It has a fixed-position elevated heating chamber with open bottom, a

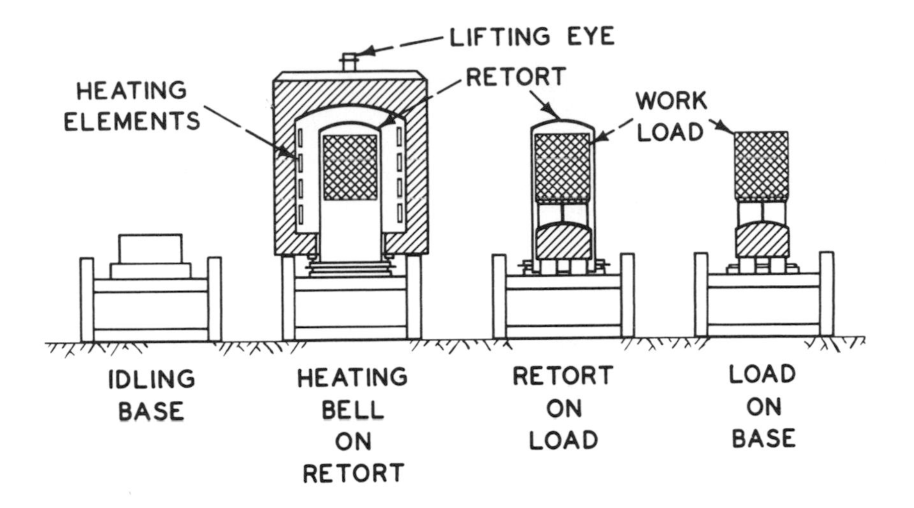

FIGURE 31-9

Bell furnace-schematic.

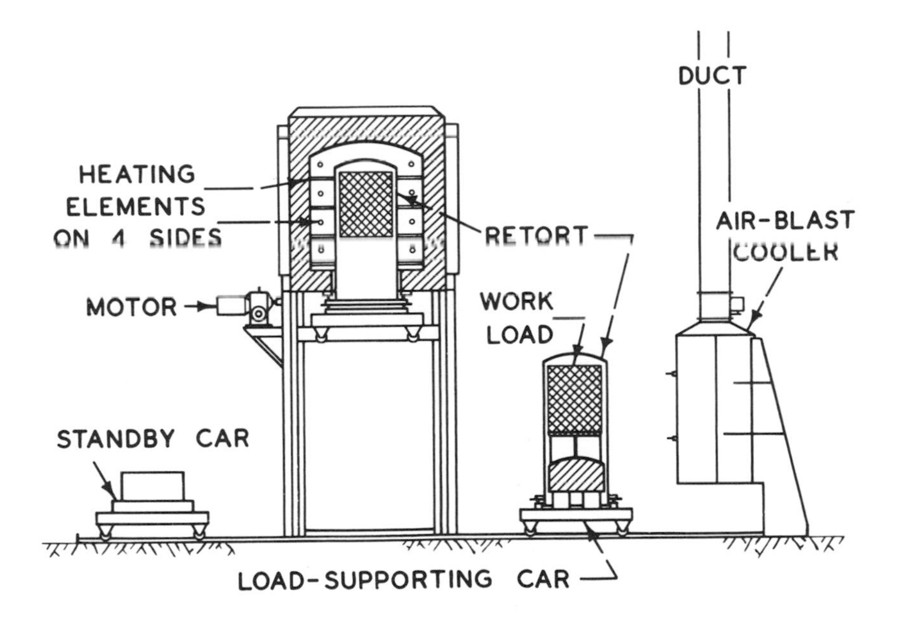

FIGURE 31-10

Elevator furnace-schematic.

mechanism for raising and lowering load-supporting cars in and out of the furnace, a standby car to plug the furnace opening during idling periods, and optional cooling chambers for hot loads from the furnace. Flexible hoses carry atmosphere gas and cooling water to and from the cars.

31.20 Vacuum Furnaces

Bell type vacuum furnaces are used for sintering refractory metals. The refractory powder, usually in the form of an electrode is pre-sintered in a hydrogen-atmosphere furnace. The electrode is then placed in the vacuum bell furnace, and current is applied to the ends of the electrode to obtain the high temperature required for complete sintering.

Tungsten and other cemented carbide parts are sintered by a liquid phase process during which the carbides are cemented or embedded in the cobalt.

Because it is relatively easy to obtain high temperatures and to control the sintering environment in vacuum furnaces, they have almost completely replaced atmosphere sintering furnaces for hardmetal processing. Higher yields, lower operating costs and better product uniformity are the result. Resistance furnaces have steadily won ground over induction furnaces on a first cost basis, particularly as work zone size has increased. With the development of special low temperature controls and wax collection methods, it is possible to remove the paraffin binder and sinter in the same furnace. Batch furnaces are used almost universally. They are often connected in pairs to a common vacuum pumping system, power supply and controls, since about 50% of the cycle is required for pulling a vacuum and heating for sintering, the balance being used for cooling.

A typical vacuum sintering furnace for cemented carbide parts is shown in figure 31-11. The work is placed on carbon trays which are then stacked and loaded into the furnace on graphite hearth plates. The plates in turn rest on graphite piers or rails. The work is heated by graphite resistance elements, which face the work on two or four sides and which are surrounded with insulation consisting of carbon felt layers within a light stainless steel structure. Enclosing this insulated box, or hot zone, is the "cold wall" vacuum containment vessel, generally a water cooled, double walled horizontal cylinder with loading doors at one or both ends. The hot zone has openings with moveable shutters which are closed during heating. The work may be cooled quickly after sintering by opening the shutters and circulating inert gas through the zone using a blower or fan mounted in or ducted to the chamber. A vacuum pumping system with wax trap, water heating and circulating equipment, a power supply and electrical controls and instrumentation completes the system.

A complete sintering cycle may take from 6 to 14 hours or even longer for

FIGURE 31-11
Batch type vacuum sintering furnace.

thick sectioned parts during which temperature and pressure profiles are generally similar to the curve in figure 31-12. Dewaxing, or binder removal, requires slow heating to prevent the expansion of vapor from fracturing the parts. The furnace temperature control system must be specifically designed for accurate control from room temperature to 660F to (350C) during this phase. After dewaxing, the furnace temperature is increased more rapidly to 1650 to 2200F (900 to 1200C), often with one or two soak plateaus to even the temperature gradients within the load. Some cycles are terminated at this point and the work cooled in the presintered form. At this stage, the parts are strong enough to be machined but are not so hard as to require diamond tools. For full sintering the temperature is brought to approximately 2675F (1470C), or higher depending on the material, and held for 15 to 30 minutes, to permit complete development of the liquid phase and the requisite grain growth. After sintering, the furnace power is turned off and the work cooled, initially in vacuum and subsequently in inert gas circulated through the hot zone. Although argon is used generally, in some instances hydrogen is substituted for faster cooling.

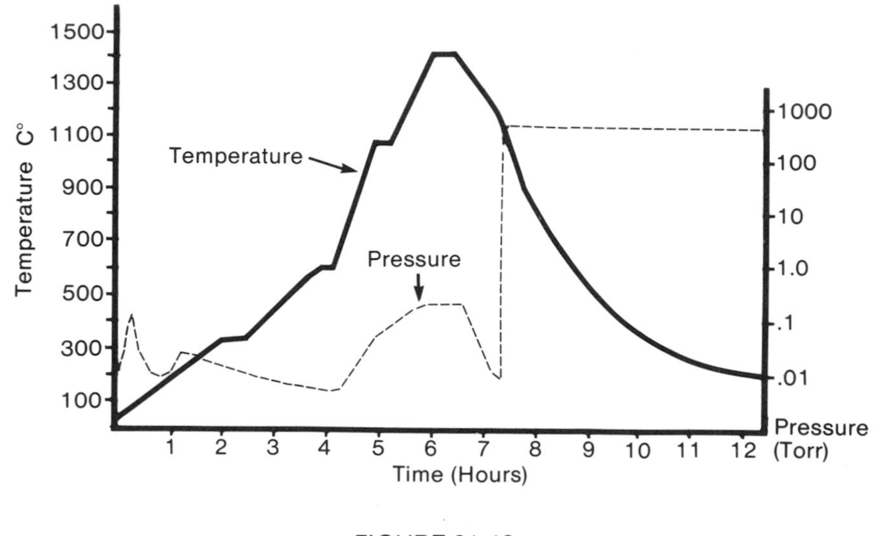

FIGURE 31-12
Typical vacuum sintering cycle.

During various phases of the cycle partial pressures of inert or active gases may be injected into the vacuum system. Argon aids dewaxing, hydrogen deoxidizes and methane adds carbon. Generally the vacuum pumping system is continually connected to the furnace chamber during the entire cycle (except during cooling) to pump away any outgassing or products of reaction. During final sintering, the pump is often throttled allowing the chamber pressure to rise enough to inhibit the vaporization of cobalt. Pumping equipment usually includes a Roots blower backed with a mechanical vacuum pump. Operating pressures range from 1 torr to 10 millitorr. Pump size is related to the types of material and the average gas content to be expected in the work.

Because excessive paraffin may destroy the lubricating effect of the vacuum pump oil, effective condensing systems are necessary to trap the vapor before it reaches the pump. Almost all systems today are equipped with microprocessor based temperature controls, acting through transformer or SCR power supplies. Accurate temperature measurement is important for process control and tungsten-rhenium or platinum-rhodium alloy couples, with ceramic or molybdenum protective sheaths are the most frequently used sensors. Some furnace systems have been built with optical temperature sensing devices and portable pyrometers are universally used as a check on the thermocouples. With the advent of microprocessors it has become practical to make equipment operation completely automatic with the exception of loading and unloading. The process cycles can be varied on a preselected basis to account for load size and material.

32.10 Sintering Zones

32.11 Burn-Off and Entrance Zones

The burn-off chamber, sometimes referred to as the heated purge chamber or entrance zone, is an essential part of the sintering furnace. Stearic acid, or various metallic stearates used as lubricants for compacting will interfere with good sintering if not expelled properly before the compacts reach sintering temperature. The compacts should be heated slowly at this stage so that the entrapped air and lubricants will not expand too rapidly and push the metal particles apart. It is very important that all of the lubricants be volatilized and expelled with the outflowing furnace atmosphere before the compacts enter the high temperature zone of the furnace. Burn-off chambers are usually controlled to heat compacts to approximately 800 to 1500F (430 to 815C). Sufficient input should be provided to heat to at least 1600F (870C), because different types of lubricants volatilize at different temperatures. The burn-off chamber must be also of sufficient length to allow the compacts to reach the temperature needed to completely volatilize the lubricant.

If the lubricants are not expelled completely before the compacts enter the high temperature sintering zone, considerable difficulty may be encountered. Since the furnace protective atmosphere is reducing, there is no air in the furnace to burn up the hydrocarbon vapors. The hydrocarbons break down into free carbon which deposits on the refractories and on the electrical heating elements. If the furnace is not shut down frequently to burn out the carbon by means of introducing air into the furnace, the heating elements will short out and fail, and the refractory will fail due to impregnation with carbon. Even if the furnace is of the full muffle type, carbon will deposit on the muffle to impair heat transfer.

When zinc stearate is used as a lubricant, the zinc as well as the stearate hydrocarbon will add to the problem of operating the furnace. If not expelled completely in the burn-off chamber the zinc will volatilize in the high temperature zone to contaminate the heating elements and the furnace alloy components in an atmosphere-tight furnace. Eventually it will work its way to the cooling chamber and condense on the sides. In the cooling chamber this will reduce the heat transfer and the compacts will not cool sufficiently. If this condition exists, the furnace must be shut down, and the cooling chamber scraped out.

Not only will the lubricants cause excessive furnace maintenance if not completely expelled before the compact enters the high temperature zone, but they can affect the quality of the product. The dimensions of the parts may vary because the lubricants are being expelled too rapidly when the parts reach the heat zone. Carbon from the lubricants may discolor the part and make it a reject. The discoloration can vary from part to part, because the lubricants may have been completely expelled from some and not from others.

The flow of the atmosphere also plays an important part in expelling the lubricant vapors when the burn-off chamber is attached to the high heat zone of the sintering furnace. Sufficient atmosphere gas must be provided, and the flow directed so the vapors are discharged toward the furnace entrance and not into the high heat zone.

Sometimes the burn-off chamber is separated from the high heat sintering chamber. The two chambers are separated with an air gap, using a flame curtain before the high heat chamber. Thus, any volatiles escaping from the burn-off chamber cannot enter the high heat chamber when a load of compacts is charged, provided the compacts have been heated sufficiently to drive off all the lubricant. The separate burn-off chambers are usually semi-muffle gas-fired furnaces. They will burn up the volatile lubricants and discharge them with the flue products.

The separate burn-off chambers are usually operated so the compacts reach a maximum temperature of 800° F (430° C). The products of combustion in general do not excessively oxidize the green compacts up to this temperature. Combustion can be adjusted toward the reducing side to keep oxidation to a minimum. However, when processing compacts with graphite additions, this type of burn-off chamber can lead to partial or total loss of graphite due to oxidation.

A considerable number of gas-fired, semi-muffle separate burn-off chambers have been installed on continuous furnace installations, and excellent sintering has been obtained with the complete absence of high furnace maintenance associated with these furnaces when using attached burn-off chambers.

If the burn-off chamber is attached to the sintering furnace, the design of the burn-off chamber's electric heating elements must be such that the hydrocarbon vapors, zinc and carbon do not affect them. This is done either using a muffle at this point, or sheathed elements or low voltage elements. As a substitute for electric elements, gas fired radiant tube heating units can be used.

Considerable attention should be given to the design of the burn-off chamber because of its importance to good sintering and low furnace maintenance, and the fact that it also serves as a first stage preheat.

Burn-off of the lubricants for vacuum sintering must be given special attention. In the first place, metallic stearates should not be used. Instead, stearic acid or a wax compound should be used. Zinc or lithium residue left behind after the burn-off of a metallic stearate will volatilize at the high sintering temperature and contaminate the furnace and vacuum pumps. For batch-type vacuum furnaces, the burn-off is best accomplished in a separate protective-atmosphere oven at about 800F (430C). An air atmosphere in an electric oven or the products of combustion in a gas-fired oven will degraphitize iron-graphite P/M parts at temperatures as low as 700F (370C). This will result in a decarburized layer after sintering. Air-atmosphere electric ovens are quite suitable for dewaxing austenitic stainless steel P/M parts at about 800F (370C). The carbon-bearing martensitic stainless steels should be dewaxed in a protective atmosphere to prevent decarburization before vacuum sintering.

Special semi-continuous furnaces have been built for the sintering of tungsten carbide P/M compacts, in which the carbide parts are dewaxed under vacuum. A cold trap is used to condense and separate the volatilized wax before it reaches the vacuum pump. The vacuum dewaxing furnace is designed so that the temperature of the walls of the furnace are maintained at about the melting point of the wax to prevent condensation of the wax on the furnace walls or heat shields. Vacuum dewaxing has several advantages over protective atmosphere dewaxing. The dewaxing takes place at a much faster rate, and vacuum is less costly than protective atmosphere and is more thorough in the removal of the wax at the low temperature.

32.12 High Temperature Zone

The high temperature zone must be of the proper length in relation to the burn-off and/or preheat zones to allow sufficient time at temperature to obtain the desired density and strength. The lengths of the burn-off and high temperature zones are usually about equal. The chief causes of poor strength or density in a powder metallurgy product is that either a sufficient temperature was not reached or that the part was not held long enough at the proper temperature to obtain good bonding of the particles. Therefore, in sintering furnaces, the high-heat chambers are made so that heat input is concentrated at the entrance end and reduced in the soaking zone.

32.13 Cooling Zone

The cooling zone of a sintering furnace often consists of a short insulated cooling zone, followed by a long water-jacketed cooling zone to cool the parts sufficiently to prevent oxidation upon their discharge into air. The length of the cooling zone is usually 2 to 3 times the length of the high temperature zone. The short, insulated cooling zone permits the parts to cool from the high sintering temperature to a lower temperature at a slower rate so as to prevent thermal shock. The insulated cooling zone also cuts down on maintenance of the long water-jacketed cooling zone, belts, trays, and fixtures by preventing the high stresses from thermal shock.

Furnaces become quite long if cooling is to be provided below 300F (150C). The cooling rate is exceedingly slow at the low temperature range. Fans sometimes are used in the cooling chambers to help cool by circulating the atmosphere over the work. Extreme care must be taken in the design when using fans, so that air is not sucked into the cooling chamber when the doors are opened.

Automatic water temperature control is essential on cooling chambers for fool-proof operation. If a cooling chamber is operated too cool, below the dew point of the furnace atmosphere, condensation will occur on the walls of the chamber, and then parts will become blue or oxidized. If the chamber is operated too hot, the parts will not cool sufficiently, and will come out of the cooling chamber hot and oxidize in the air. To compensate automatically for varying loads, automatic flow control of the water by means of a thermostat and throttling control valve is the only foolproof method of operation.

33.00 METHODS OF HEATING

33.10 Fuel-Fired Furnaces

Fuel-fired furnaces are either direct-fired muffle, or radiant-tube-fired. The reason for this is the necessity for isolating the work and its protective atmosphere from the products of combustion. Limited use is made of semi-muffle furnaces and at presintering temperatures only to remove lubricants from the compacts. Fuel-fired furnaces are commonly applied to the sintering process where fuel economy and low-range temperature are dominant factors.

Uniformity of heating and close control of atmosphere conditions are critical factors in the sintering of metal powders. Both the direct-fired muffle and radiant-tube-fired furnaces meet these requirements.

33.11 Muffle-Type

High temperature operation requires a material to withstand high temperatures, and to be capable of withstanding the thermal shock which results from pushing the work through the furnace. Refractories meet the temperature requirements, but because of the thermal conductivity and thermal shock properties of refractories, cracks may develop rather easily, joints may not be tight, and the refractory may be porous. These openings allow products of combustion and/or air to leak in to contaminate the atmosphere. Therefore the use of metallic alloy muffles normally is a definite requirement for fuel fired furnaces.

Since very dry, high-purity atmosphere is a requirement for sintering stainless steel parts, metallic muffles are almost universally used in the application. The most popular shapes of muffles have "D" or circular sections, often supported fully in their lengths by refractory hearth plates. A furnace with a cylindrical muffle on refractory piers is shown in figure 33-1. Muffles require the proper strength to minimize sagging, bulging, warping, and cracking at high temperatures.

To ensure good atmosphere conditions during normal operation and an oxide-free surface on the inside of the muffle, a reducing atmosphere must be maintained when the muffle reaches the red heat range. On shut downs, the atmosphere must be maintained until the black heat range is reached.

Muffle furnaces are most satisfactory for low production service, and where ultra-high purity and conservation of atmosphere is an important factor.

33.12 Radiant-Tube Type

Radiant tubes are used to heat high-production sintering furnaces. These furnaces do not require an alloy muffle, but have an atmosphere-tight refractory-lined casing. They have the advantage of less maintenance and conservation of alloy. A typical cross section of a vertically-fired radiant tube furnace is shown in figure 33-2. Much sintering work can be done in atmospheres that are not of the highest purity and atmosphere-tight radiant tube

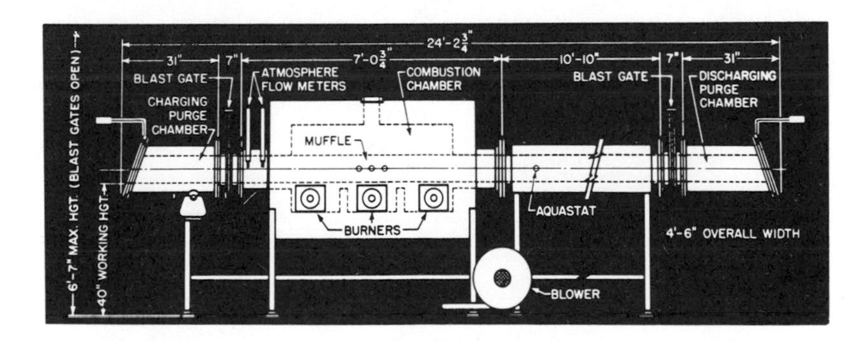

FIGURE 33-1
Longitudinal section of muffle furnace.

FIGURE 33-2
Interior of vertical radiant tube furnace.

fired furnaces meet these requirements satisfactorily. Although better purging and less atmosphere are advantages of the muffle furnace, furnace size limitations restrict muffle usage. Radiant tube fired furnaces permit low maintenance operation up to 1850F (1010C) when using nickel-chromium alloy tubes, and up to 2000F (1093C) when using super alloys or ceramic-coated tubes. Because of the temperature limitation, the most common applications of radiant tubes in sintering furnaces are in burn-off chambers.

33.13 Economy of Fuel-Fired Furnaces

Economical operation of the horizontally- and vertically-fired radiant tube furnaces is limited to operations below 1850F (1010C). Above this temperature, stack losses are excessive as shown in the Table 33-I below.

TABLE 33-I

Exhaust Gas Temp. F	C	Stack Loss	Available Heat % Gross Input
400	204	16	84
600	316	21	79
800	427	26	74
1000	538	30	70
1200	649	35	65
1400	760	40	60
2000	871	45	55
1800	982	50	50
2000	1093	56	44
2400	1316	67	33
28	1538	80	20

Also, operation above 1850F (1010C) leads to high maintenance costs of the alloy radiant tubes. Because of this, electric furnaces generally are used for higher temperature sintering. An exception is noted in the case of the small diameter alloy muffle furnaces which can be operated successfully up to 2100F (1149C).

33.14 Semi-Muffle Type Furnaces

The work chamber of fuel-fired semi-muffle furnaces is not isolated from the products of combustion, and no attempt is made to use atmosphere in these furnaces. The muffle protects the furnace load from direct flame impingement even though the products of combustion can and do enter the muffle.

This type of furnace is used to burn off the volatile matter in the compacts prior to sintering. To avoid or minimize oxidation, the furnace is used to heat the work to 800F (425C) maximum. The products of combustion and the volatiles are discharged together through the semi-muffle vent stack.

When determining whether to use fuel-fired furnaces or electric furnaces, fuel costs are not the only factor to consider. If a general statement could be made regarding a choice, it might be this: the choice of a fuel fired furnace or an electrically heated furnace should be predicated on the cost and availability of alternate heating sources. However, burn-off chambers can be either gas-fired or electrically heated. To select a gas-fired furnace merely because gas happens to be cheap in the particular area of the furnace operation might be unwise, unless engineering calculations have been made to determine the efficiency of heat release through the muffle to the work. As noted previously and shown by the heat-loss table, the stack losses mount rapidly with increasing temperature of operation. The restriction of heat release through the muffle or radiant tube to the work is another big factor.

When using high-purity hydrogen and dissociated ammonia in small gas-fired furnaces, alloy-muffle types are most successful in maintaining the high-purity atmosphere free from contamination. Muffles sometimes are used in electric furnaces where extremely dry hydrogen atmospheres are required.

33.20 Electrically-Heated Furnaces

There are three basic kinds of electric heating elements used in sintering furnaces: base-metal nickel-chromium alloys, non-metallic heating elements, (silicon-carbide or graphite) and refractory-metal heating elements (molybdenum or tungsten). Graphite and refractory metal heating elements are used almost exclusively in vacuum furnaces.

All three types have definite and specific applications to sintering. There can be some overlapping but the wrong application, particularly with reference to furnace atmosphere, can cause very high furnace maintenance. Since the furnace atmosphere is so important to the life of the heating elements, this subject will be dealt with in more detail later.

33.21 Nickel-Chromium Heating Elements

Nickel-chromium alloy heating elements are used in electric furnaces for heavy duty continuous operation up to 2100F (1150C) in any type of furnace atmosphere except strongly carburizing or sooting atmospheres. The 80% nickel-20% chromium alloy generally is used for the higher temperature ranges. The 35% nickel-20% chromium alloy is suitable up to 1800F (980C) operating temperature in protective atmosphere. There are some cases of contamination of furnace atmospheres, to be discussed later, where the 35% nickel-20% chromium alloy is more suitable than the 80% nickel-20% chromium alloy. Heating elements are generally designed in solid rod overbend or flat ribbon overbend loops, and are supported on the refractory walls of the furnace by means of alloy pins, hanger bolts, or refractory supports as shown in figure 33-3. Also, they can be mounted in the furnace hearth and supported from the furnace roof or arch. Elements can be used to distribute the heat evenly, or concentrated in areas where most needed. Some element mountings are built-in, while others are removable. The elements are strong and ductile when either cold or hot.

The watt density loading of element surfaces, play a very important part in the life of the alloy heating element. The higher the operating temperature, the lower should be the watt density loading. For example, if the heating element is to operate at temperatures in excess of 1750F (954C), the watt

FIGURE 33-3
Overbend heating elements supported on furnace wall.

density should be in the order of 8 to 12 w/ in^2 surface area (1.24 to 1.86 w/cm^2). For temperatures under 1750F (955C), the watt density may go as high as 17 w/ in^2 (2.64 w/cm^2), depending on the operating temperature and whether the furnace is to operate continuously with heavy production loads, or intermittently with light production loads.

The cross-section of the alloy plays a very important part in the life of the heating element. Since most reducing atmospheres adversely affect the life of the element, the cross-section of the alloy should be as heavy as practical for a given design and input. The reason for this is that the attack by the atmosphere at the surface will cause a great change in the resistance of a thin cross-section, but will have only a small percentage of change on a heavy cross-section in the same period of time.

Heavy-cross-section heating elements not only cost more than the thin-section elements, but also may require transformers because the heating element voltage often is well below the normal 230 or 460 volt supply. This extra initial expense is a good investment, because the maintenance expense when using the heavier-cross-section elements will be reduced greatly.

Another good reason for operating heating elements of any type on low voltage is to minimize shorting and arcing between the element bends and leads, due to carbon deposition from the atmosphere, or lubricants that may carry over into the high-temperature sintering chamber. Under favorable conditions of design, atmosphere, temperature and care, nickel-chromium elements have long life and low maintenance cost.

33.22 Non-Metallic Heating Elements

(a) Silicon Carbide: Non-metallic, silicon carbide heating elements are made in rod form. The ends of the rods extending through the insulation of the furnace have lower resistance, so that only the portion in the heating chamber get hot. The rods usually run horizontally under and over the hearth as shown in figure 33-4. However, it is also possible to run the rods vertically. The electrical connection is made by aluminum braid fastened to the rod by a chromium-steel spring clip, and the other end of the braid bolted to the bus bar, as shown. The terminal box on an atmosphere furnace is sealed with a cover plate and gasket. The silicon-carbide bars have low tensile strength when cold or hot and cannot withstand shock or high stress. They must be able to expand and contract freely as their temperature changes.

These elements can be replaced quickly and easily. In fact, the bars can be replaced while the furnace is hot. This is an important factor in large continuous furnaces which would require days to cool to get inside and replace built-in heating elements.

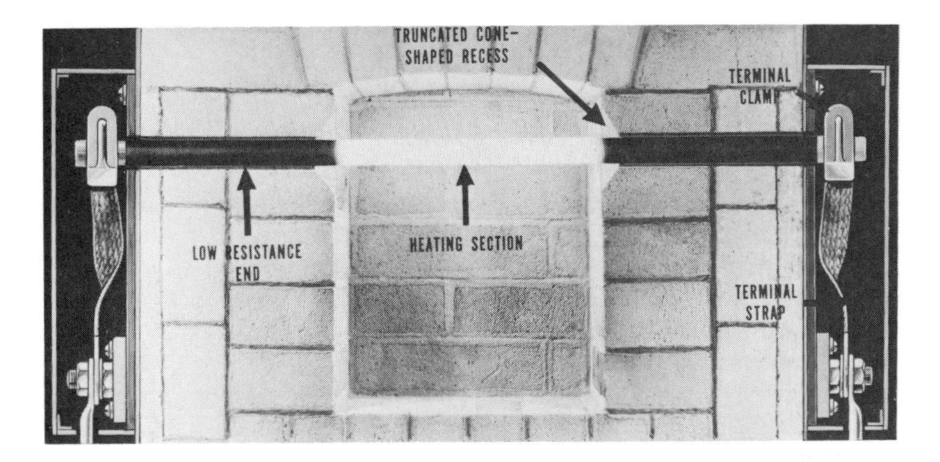

FIGURE 33-4

Method of installing silicon carbide heating elements.

This type of element is particularly useful in sintering furnaces operating from 1850F to 2100F (1010 to 1150C) in non-decarburizing atmospheres and in plants which require a furnace to be universally used with high or low carbon potential atmospheres on both ferrous and nonferrous parts. Reducing atmospheres adversely affect their life above, but not below, 2425F (1330C) element temperature.

A conservative watt density loading on silicon-carbide elements is 30 w/in^2 (4.65 w/cm^2) in reducing atmospheres at 2100F (1150C). Higher loadings may be used when the elements operate in air. The high watt density loading of the silicon-carbide element is another advantage over the nickel-chromium type. More power input can be concentrated in the furnace and, therefore, reduce the wall area and overall size of the furnace.

Silicon-carbide elements increase in resistance on aging, and variable-voltage transformers or saturable core reactors are normally required. This type of transformer is more costly than the constant-voltage type used with nickel-chromium elements. This is one disadvantage of silicon-carbide resistors and usually the initial cost of the overall equipment installation is greater because of it. Other disadvantages are that they do not all age at the same rate, particularly if connected in series-parallel circuits, necessitating matching resistance upon replacement. They are brittle and fragile at all temperatures, and have higher thermal losses due to the terminal holes in the walls of the furnace. However, the advantages more than outweigh the disadvantages, and silicon-carbide elements are used widely in sintering furnaces.

(b) Graphite: Graphite heating elements in the form of tubes, rods, plates or cloth are used extensively in vacuum sintering furnaces. Since vacuum furnaces operate in the absence of air or oxidizing gases, graphite is a suitable heating element material for temperatures up to 4000F (2200C). Graphite is relatively low in cost and is the only material that increases in strength as the operating temperature increases. Once graphite has been heated and outgassed in a vacuum furnace, it no longer gives off any carbonaceous gases that can be carburizing to the P/M parts during sintering.

33.23 Refractory-Metal Elements

(a) Molybdenum: Molybdenum elements often overlap or take over from silicon-carbide elements in the high temperature range. There are two element designs used in sintering furnaces:

In the small-size furnaces, the molybdenum elements are wound directly onto alumina muffles of round or "D" shape. Coatings of cement are then applied over the windings to hold them in place. The muffle is placed in a gas-tight welded shell and insulated with a high purity alumina refractory. The leads from the elements must be sealed by means of packing glands. This design is used chiefly on laboratory furnaces having muffles from 3 to 6 in (7.6 to 15.2 cm) in diameter or width. The construction is not suitable for larger furnaces because the refractory will sag at the high temperature and break the element.

In larger furnaces the design is identical to nickel-chromium design of rod overbend or ribbon overbend elements. The construction allows direct radiation of heat to the work. The elements are supported on the walls of the furnace by molybdenum pins or hanger bolts, or by refractory supports. The furnace may be lined with additional insulation for the higher-temperature operation. If a dry-hydrogen atmosphere is to be maintained, the refractory must be a high-purity alumina brick to prevent deterioration in the purity of hydrogen atmosphere. Molybdenum and high purity alumina refractory are expensive. Therefore, a molybdenum element high temperature sintering furnace can be a very expensive item when compared to the same size nickel-chromium or silicon-carbide heated furnaces. On the other hand, molybdenum elements properly applied and cared for do not deteriorate as do the others. Barring accidents, they may last many years and, therefore, can have the lowest maintenance cost of all.

Molybdenum has the characteristic of becoming coarse-grained and quite brittle after it is heated, more so at room temperature than at high temperature. Therefore, care must be taken not to thermally shock the element by too great a change in electrical input, or by charging a load of cold work directly into the furnace without a preheat. The resistance of molybdenum is much lower at room temperature than at elevated operating furnace temperatures so starting taps of lower voltage are used to prevent thermal shock or overloading. The transformer, however, is not as expensive as the silicon-carbide-element transformers which must have many voltage taps. Saturable-core reactor control is often preferred to an on-off control to provide unattended, safe start-up for the molybdenum heating elements. Also, by increasing the voltage on the elements as the temperatures and resistance increase, while holding constant current, the power input increases. This is a desirable feature, since more energy usually is needed at higher temperatures.

Molybdenum elements can be operated up to 3400F (1870C) element temperature in a hydrogen atmosphere, but consideration must be given to the use of proper refractories. The watt density loading for molybdenum can be as high as 50 to 60 w/in^2 (7.75 to 9.30 w/cm^2). This allows considerable input into a small chamber to overcome radiation losses which is required for high temperature operation. In designing a molybdenum element, the current density (amperes per in^2 or cm^2 of cross-section) must also be taken into the design calculations, as well as the watt density loading.

Molybdenum elements can be welded and spliced, but the welds are brittle and must be handled with great care. Warming up helps their ductility.

Where possible, the element is usually made of one continuous piece of molybdenum wire. The lead-out terminals are made to have low resistance by twisting several extra strands of molybdenum wire around the terminals so excessive temperature will not be developed in them.

Molybdenum heated furnaces are used for sintering stainless steels, refractory metals and cermets; carburizing tungsten to make tungsten carbide, and sintering tungsten carbide to make tool and cutting tips.

(b) Tungsten: Tungsten heating elements are used specifically in vacuum sintering furnaces above the top limit for molybdenum heating elements. Tungsten is more expensive and more difficult to form than molybdenum and is used only for vacuum furnaces at temperature ranges from 3300 to 4500F (1815 to 2480C) to sinter refractory metals and carbides.

34.00 TEMPERATURE CONTROLS

Instrumentation plays an important part in temperature control required for good sintering. On-off control and proportioning controls are two widely used types. In connection with either of these, the proper sensing device (usually a thermocouple) must be used.

In addition to the control thermal element, some type of thermocouple protective device should always be installed to prevent run-away furnace temperature in the event of thermocouple failure. However over-instrumentation can lead to high maintenance costs.

34.11 On-Off Control

The primary temperature-sensing element is a thermocouple extending through the furnace roof or wall into the heating chamber (figure 34-1). A difference in temperature between the hot end and cold end of the thermocouple induces a millivolt output from the thermocouple, which varies with temperature. This electric signal is conducted to a temperature-control instrument, often called a "pyrometer". Pyrometer controls operate a contactor for on-off power input or actuate a motor or solenoid operated valve to regulate the input of combustion air. Long or high furnaces often have several thermocouples and "zones" of energy control. Zoning helps apply high heat in heating-up areas, and moderate heat input in soaking zones.

34.12 Proportioning Control

Proportioning controllers are of the position-adjusting, duration-adjusting or current-adjusting type. The position-adjusting type is used with gas-fired furnaces. The duration and current-adjusting types are used on electric

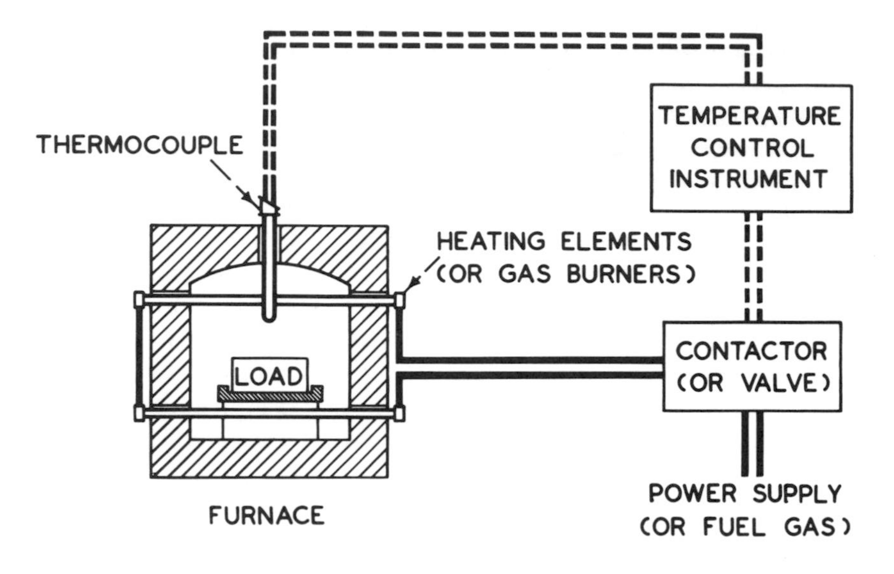

FIGURE 34-1

Typical on-off temperature control system.

furnaces. The controllers develop proportioning type of control by incorporating proportional, reset and rate actions. These three actions regulate temperature closely by continuously monitoring heat input according to the speed, degree and duration of temperature changes. Proportioning control is used in conjunction with a standard pyrometer whenever straight line temperature control is required.

a) Position adjusting controls

Proportional action adjusts the valve opening to balance the heat supply required to maintain the desired temperature. This eliminates cycling, but without a reset action will allow the temperature to wander with load changes.

Reset action adjusts the heat supply in proportion to length of time the temperature is away from desired value. This eliminates offset after a load change by continuously adjusting heat supply.

Rate action changes the input rate of heat supply in proportion to the speeds at which the change is occurring. This shortens the time needed to return the process to the desired temperature and eliminates or reduces overshoot.

b) Duration adjusting control

Duration-adjusting proportional controllers adjust the proportional reset and rate actions of heat input by regulating the successive durations of on-time to off-time according to the size, speed and duration of temperature changes.

c) Current adjusting control

With the current adjusting control, the three control functions are the same as described under positioning-type control except that a magnetic amplifier and reactor replaces the valve.

A saturable-core reactor is essentially a set of coils wound on a transformer core having a variable impedence to regulate current flow. (This could be called an electricity valve.) The reactor is connected in series with the heating element load and by means of direct current in the control winding, varies the current in the load.

With the addition of a standard temperature measuring instrument and a current-adjusting type control feeding into a magnetic amplifier, the furnace can be controlled to a very narrow temperature band.

An advantage of a saturable-core-reactor type control is that power input to the furnace will match the demand. Also, the stepless-control feature of a reactor prolongs element life. Overheating is prevented because the system maintains the lowest possible temperature gradient between the furnace and material being heated.

When applying a saturable-core reactor to a molybdenum furnace a current-limiting system should be used to prevent overloading the heating element when starting a cold furnace because the cold resistance is only one-fifth to one-tenth the hot resistance. The automatic current limiter provides overload protection for the molybdenum. A schematic diagram of this system is shown in figure 34-2. When full load current is exceeded, the current signals the magnetic amplifier to maintain the load current at approximately its full load values. This brings the furnace up to temperature automatically and unattended in a safe and rapid manner.

A disadvantage of a saturable-core control system is the higher initial cost of the equipment compared to the cost of on-off control or a duration-adjusting control. The saturable-core reactor with a current-limiting system is desirable on molybdenum furnaces as previously stated, to provide fast start-up or fast heat cycling of a furnace. Saturable-core reactor systems are used also on other types of heating elements because of the better temperature control, and the elimination of contactors. This system can also increase heating element life materially.

34.13 Thermocouples

The number of metals and alloys suitable for use as thermocouples on a commercial basis is limited. At high temperature the materials must be

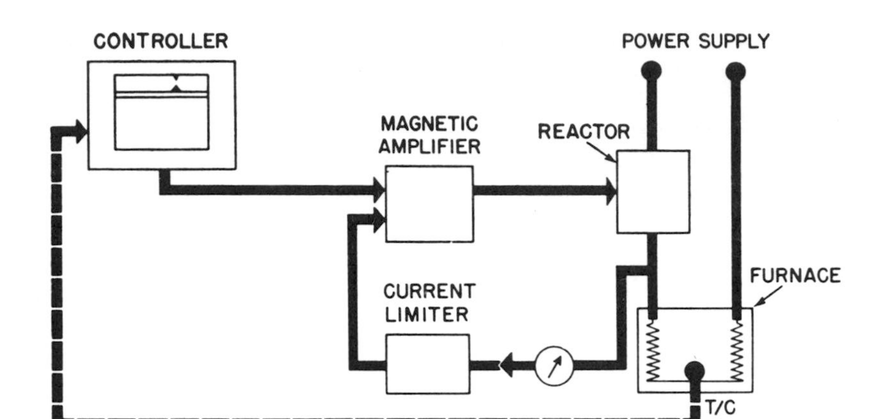

FIGURE 34-2

Typical current limiting temperature control system.

resistant to oxidation, recrystallization, melting, and contamination by reducing atmospheres. They must develop an e.m.f. high enough to be measured without the use of delicate instruments, be of reasonable cost, be readily obtainable in uniform quality, and be reproducible.

There is a temperature limit for different commercially available thermocouples in different atmospheres. It is important, therefore, that the proper type couple be used for the expected operating temperature range (Table 34-I).

TABLE 34-I
Type of Thermocouples

Type	Temperature Range
Copper - Constantan	-300° +600° F (185° to +315° C)
Iron - Constantan	0° +1400° F (-18° to 760° C)
Chromel - Alumel	0° to 2000° F (-18° to 1095° C)
Nickel-Nickel Moly	0° to 2400° F (-18° to 1315° C)
Platinum-Platinum Rhodium	0° to 2800° F (-18° to 1540° C)
Tungsten-Tungsten Rhenium	1000° to 4300° F (+548° to 2370° C)

In sintering temperature ranges up to 2000F (1095C) for nonferrous and ferrous applications, chromel-alumel thermocouples are used. Also, they may be used occasionally up to 2400F (1315C). In applications where sintering is done above 2000F (1095C) for iroin P/M parts, radiation pyrometers and platinum-platinum rhodium thermocouples usually are used.

Above 2500F (1370C) radiation-type detectors or tungsten-tungsten rhenium thermocouples are generally used. Bare nickel-nickel moly thermocouples have been used with some degree of success up to temperatures of 2400F (1315C) in vacuum furnaces. Nickel-nickel moly is the only known thermocouple material that will withstand a vacuum environment without a protection tube.

a) Chromel-Alumel

The chromel-alumel couple uses chromel as the positive element and alumel as the negative element. Chromel is an alloy composed of 90% nickel and 10% chromium. Alumel is an alloy of 94% nickel, 2% aluminum, 3% manganese, and 1% silicon.

The recommended wire sizes, B and S gauge, for chromel-alumel couples in sintering furnaces, are 8 and 14 gauge. Thermocouple response is faster with the lighter gauge wire; however, the heavier wire resists deterioration.

Chromel-alumel couples have been found to be reasonably stable thermoelectrically when heated in a clean oxidizing atmosphere. When the couple is exposed to a reducing atmosphere containing wet hydrogen or carbon monoxide, it becomes contaminated, and the output is altered. It is, therefore, necessary to protect the thermocouple from the reducing gases. This is done by use of a protection tube, vented with circulating air, nitrogen, argon or helium. Thermocouple protection tubes can develop leaks, and therefore couples should be checked frequently. Furthermore it is wise to install a second thermocouple connected to an excess-temperature cut-off as a safety measure.

Protection tubes for chromel-alumel thermocouples for temperatures up to 2000F (1095C) are made of nickel-chrome or nickel-chrome-iron heat-resisting alloys. In clean atmospheres such as hydrogen or dissociated ammonia protective thin-wall tubing of 80 nickel-20 chrome usually is used. Where the atmosphere is carburizing, as in low dew point endothermic gas with or without enrichment, protection tubes of heavy-wall Inconel pipe are used. Although the response to temperature change is slower, longer life of the tube is a compensating factor. Protection tubes of lower alloys such as 35-15 and 18-8 Ni-Cr are being used in applications where atmosphere is clean and temperatures are in the lower range.

For vacuum applications up to 2000F (1095C) chromel-alumel thermocouples generally are embedded in magnesium or aluminum oxide in Inconel sheaths.

b) Nickel-Nickel Moly

Bare nickel-nickel moly thermocouples have been used with some success in vacuum furnaces up to 2400F (1315C). This same type of thermocouple has been used with the proper protective-sheath material sealing the thermocouple from the furnace atmosphere and air for temperatures up to 2400F (1316C). The source for this thermocouple wire has been limited and for this reason nickel-nickel-moly thermocouples are not used widely, although the thermocouples are less costly than platinum-platinum rhodium.

c) Platinum-Platinum Rhodium

Platinum-platinum rhodium couples are called noble-metal thermocouples. They are made of pure platinum negative elements, and the positive element is either 87% platinum, -13% rhodium or 90% platinum, -10% rhodium.

"Platinum" thermocouples are easily contaminated. Hydrogen is especially detrimental to platinum, and will ruin a couple after even short exposure above 2000F (1095C), and even at lower temperatures, though at a slower rate. Silicon and other metallic vapors will also contaminate the "platinum" thermocouple, and therefore, protection tubes are always required. The couples usually are protected with a double tube. First, multiple ceramic insulators are used over the wires. This assembly is then inserted into a small diameter ceramic thin-wall tube; this tube then is put into a large-diameter ceramic tube. The material used for protection tubes is a highly refractory porcelain which should be glazed to be impervious to gases.

"Platinum" thermocouples normally are limited to use in the temperature range from 1000 to 2500F (540 to 1370C). They can be used up to 2800F (1540C) with shorter life. They are commonly used in sintering furnaces operating in the range from 1850 to 2100F (1010 to 1150C) for iron P/M parts. Above 2000F (1095C) the inner protection tube should be vented with air, nitrogen, argon or helium to minimize hydrogen contamination. "Platinum" couples usually are not used for temperatures under 1800F (980C) because the use of double ceramic protection tubes causes sluggish control response in the lower-temperature range.

For vacuum applications up to 2800F (1540C), platinum-platinum rhodium thermocouples generally are embedded in magnesium or aluminum oxide in either molybdenum or tantalum sheaths.

d) Tungsten-Tungsten Rhenium

Tungsten-tungsten rhenium thermocouples are available now for use in hydrogen atmospheres or vacuum, with proper sheath material, up to temperatures of 4300F (2370C). Either tungsten or tantalum can be used for sheath material in vacuum. Tungsten must be used as the sheath material in a hydrogen atmosphere because tantalum becomes embrittled with hydrogen. The sheath material must be sealed to prevent air from entering at the terminals.

e) Radiation-Type Temperature Detector

The radiation-type temperature detector uses the principle of the Stefan-Boltzmann law of radiant energy, which states that the intensity of radiant heat emitted from the surface of a body increases as the fourth power of its absolute temperature.

Radiation-type detectors are used with high-temperature sintering furnaces, such as the molybdenum-element type with hydrogen or dissociated ammonia atmospheres, because they measure without coming into direct contact with the atmosphere or the high temperature. These units sight on closed-end target tubes or into the furnace to read temperature. They are used sometimes above 2000F (1095C) and commonly above 2500F (1370C).

35.00 ATMOSPHERES AND ATMOSPHERE PRODUCERS

Furnace protective atmospheres reduce oxides, prevent the formation of oxides, and/or prevent decarburization of P/M compacts during sintering. Protective atmospheres play a very important part in the field of powder metallurgy. Following is a discussion of the most common types of gases used for protective atmospheres, properties of the gases, how to purify the gases, advantages and disadvantages, applications, how the gases are made, and how the gas compositions or impurities are determined. Gas compositions and costs are summarized in Table 35-I and applications in Table 35-II.

The most widely used protective atmospheres for sintering are hydrogen, dissociated ammonia, rich exothermic gas, purified rich exothermic gas, dry or wet endothermic gas, nitrogen based blended gas and vacuum. The factors involved in the selection of the proper atmosphere tie in with the appearance, properties, and cost of the sintered compacts.

All six of these gases and vacuum have properties which help reduce oxides in powder particles and promote maximum sintering. They also prevent oxidation while assuring good machinability and good surface appearance.

Atmospheres, such as purified, rich exothermic gas, dry endothermic gas, nitrogen based blended gas and vacuum are desirable for carbon-steel compacts. They tend to minimize decarburization of the parts which would result in soft wearing surfaces and low physical properties after heat treatment.

Once the most suitable protective atmosphere has been selected, the quality of the protective atmosphere must be maintained for optimum mechanical and physical properties, with minimum discoloration and decarburization of the work pieces.

Nonferrous compacts of copper and bronze may be oxidized throughout and scaled or discolored by oxygen. They are not adversely affected by hydrogen and carbon monoxide.

Brass compacts are adversely affected by carbon dioxide and by oxygen, sulphur and water vapor, due to selective attack on zinc. This impairs mechanical and physical properties, as well as appearance. When sintered in an open furnace, brass compacts generally are treated only in pure, dry

TABLE 35-I
Approximate Compositions and Costs of Atmosphere Gases

Atmosphere Gases	Approximate Air: Natural Gas Ratio	Dew Point °F Entering Furnace	Properties At Elevated Temperatures	Carbon Dioxide CO_2	Oxygen O_2[i]	Monooxide CO	Hydrogen H_2	Methane CH_4	Nitrogen N_2	Argon A	Helium He	Relating Cost Per Unit Volume[b]
1. Hydrogen												
A. By Electrolysis of Water												
1. Direct from Cells												
a. Unpurified-Saturated		+90 to +90	R, r, D3	0.0	0.2	0.0	99.8	0.0	0.0	0.0	0.0	7.2-12.3
b. Purified[e]		<-80[k]	R, r, *N*	0.0	0.0	0.0	100.0	0.0	0.0	0.0	0.0	7.8-12.8
B. Bulk compressed gas		<-90	R, r, *N*	0.0	0.0	0.0	99.95	0.0	0.0	0.0	0.0	12-42[h]
C. By Catalytic Conversion of Hydrocarbons[c]												
1. Unpurified-Saturated		+70 to +90	R, r, D3	0.05	0.0	0.02	99.73	0.20	0.0[d]	0.0[d]	0.0[d]	8-13
2. Dried		<-100	R, r, *N*	0.002	0.0	0.02	99.97	0.01	0.0[d]	0.0[d]		
D. From Liquid Hydrogen		<-90	R, r, *N*	0.0	0.0	0.0	99.998	0.0	0.0	0.0	0.0	12-42[n]
2. Nitrogen Based Enriched With[i]												
A. Endothermic Gas		-10 to +5	R, r, C_1	0.06	0.0	4.0	008.0	0.0[1]	88.0	0.0	0.0	1.5-6.0[h]
B. Hydrogen		<-90	R, R, N	0.0	0.0	0.0	5.0	0.8[1]	95.0	0.0	0.0	1.7-7.0[h]
C. Methanol		-10 to 0	R, R, C_1	0.05	0.0	4.0	8.0	0.01[1]	88.0	0.0	0.0	1.6-6.5[h]
3. Ammonia-Based[i]												
A. Dissociated Ammonia												
1. As Reacted—Dry		-60 to -40	R, r, *N*	0.0	0.0	0.0	75.0	0.0	25.0	0.0	0.0	3.3-7.2[h]
2. Moisture Added—Saturated		+70 to +100	R, r, D3	0.0	0.0	0.0	75.0	0.00	25.0	0.0	0.0	3.4-7.4[h])
B. Burned Dissociated Ammonia												
1. Rich—Saturated[f]		+70 to +90	R, r, D3	0.0	0.0	0.0	24.0	0.0	76.0	0.0	0.0	2.3-5.1[h]
2. Lean—Saturated[g]		+70 to +90	O, n, D3	0.0	0.0	0.0	1.0	0.0	99.0	0.0	0.0	1.9-4.2[h]
C. Direct Catalytic Conversion of Ammonia and Air												
1. Rich												
a. As Related—Saturated		+70 to +90	R, r, D3	0.0	0.0	0.0	25.0	0.0	75.0	0.0	0.0	2.3-5.1[h]
b. Cooled to 40° F		+40	R, r, D2	0.0	0.0	0.0	25.0	0.0	25.0	0.0	0.0	2.3-5.1[h]
c. Dried[e]		<-80[k]	R, r, *N*	0.0	0.0	0.0	25.0	0.0	75.0	0.0	0.0	2.3-5.2[h]
2. Lean—Saturated	10.25:1	+70 to +90	O, n, D3	11.5	0.0	0.7	0.7	0.0	87.1	0.0	0.0	1.2
a. As Related—Saturated		+70 to +90	O, n, D3	0.0	0.0	0.0	1.0	0.0	99.0	0.0	0.0	1.8-4.2[h]
b. Cooled to 40° F		+40	O, n, D2	0.0	0.0	0.0	1.0	0.0	99.0	0.0	0.0	1.8-4.2[h]
c. Dried[e]		<-80[k]	N, n, *N*	0.0	0.0	0.0	1.0	0.0	99.0	0.0	0.0	1.8-4.3[h]
4. Reformed Hydrocarbon Gases												
A. Exothermic Gas												
1. Rich												
a. As Related—Saturated	6:1	+70 to +90	R, r, D3	5.0	0.0	10.0	14.0	1.0	70.0	0.0	0.0	1.0
b. Refrigerated	6:1	+40 to +60	R, r, D2	5.0	0.0	10.0	14.0	1.0	70.0	0.0	0.0	1.2
B. Purified Exothermic Gas												
1. Rich	6:1	<-40	R, r, Cl	0.0	0.0	10.8	15.0	1.0	73.2	0.0	0.0	1.5-7.2
2. Medium Rich												
a. As Reacted	6.75:1	<-40	R, r, Cl	0.0	0.0	9.0	12.0	0.2	78.8	0.0	0.0	1.5-2.2
b. Methane Added[j]	6.75:1	<-40	R, r, C3	0.0	0.0	8.5	11.4	5.2	74.9	0.0	0.0	1.7-2.4
3. Lean	10.25:1	<-40	N, n, *N*	0.0	0.0	0.7	0.7	0.0	98.6	0.0	0.0	1.8-2.5
C. Endothermic Gas												
1. Rich—Dry	2.4:1	-10 to +10	R, r, C3	0.0	0.0	20.0	38.0	0.5	41.5	0.0	0.0	1.8-2.5
2. Rich Fairly Dry												
a. As Reacted	2.6:1	+20 to +30	R, r, C2	0.0	0.0	19.0	37.0	0.3	43.7	0.0	0.0	1.5-3.1
b. Methane Added[j]	2.6:1	+20 to +30	R, r, C3	0.0	0.0	18.1	35.2	5.3	41.4	0.0	0.0	1.6-3.2
3. Medium Rich—Saturated	3.5:1	+70 to +90	R, r, D1	1.2	0.0	16.5	27.6	0.0	54.7	0.0	0.0	1.5-3.0
4. Lean—Saturated	4.5:1	+70 to +90	R, r, D3	3.0	0.0	13.8	21.5	0.0	61.7	0.0	0.0	1.5-2.5
5. Argon—Bottled[j]		<-90	N, n, *N*	0.0	0.0	0.0	0.0	0.0	0.0	88.997	0.0	28-65[h]
6. Helium—Bottled[i]		<-90	N, n, *N*	0.0	0.0	0.0	0.0	0.0	0.0	0.0	99.995	70-95[h]
7. Vacuum (Below 150 Microns)		—	R, r, *N*	0.0	0.0	0.0	0.0	0.0	0.0	0.0	0.0	—
8. Air												
A. Normal		<+90	O, o, D3	0.0	21.0	0.0	0.0	0.0	78.1	0.9	0.0	—
B. Wet		>+50	O, o, D3	0.0	21.0	0.0	0.0	0.0	78.1	0.9	0.0	—

(a) Properties at elevated temperatures with:

Iron or iron oxides:	O-Oxidizing	Iron or iron-carbon:	Cl-Mildly carburizing
	N-Neutral		C2-Adverage carburizing
	R-Reducing		C3-Strongly carburizing
Copper or copper oxides:	o-Oxidizing		*N*-Neutral
	n-Neutral		D1-Mildly carburizing
	r-Reducing		D2-Average carburizing
			D3-Strongly carburizing

(b) Costs are relative to rich exothermic = 1.0 and are based on the following data as of June, 1985:

Power	2.5¢-5¢/KWH
Natural Gas	$3.00-$5.00/1,000 CF
Water	$1.00/1,000 gallons
Ammonia	20¢/lb[h]
Steam	$3.00/1,000 lb.
Labor (including overhead)	$20-$25
Interest'	13.5%
Depreciation	10 yr, straight line
Floor Space	$75/ft²/yr

(continued)

(c) Composition varies with purification stages utilized.
C_1 Up to 1% methane or other hydrocarbon may be added for increased carbon potential.
(d) Some traces will be present if contained in fuel gas processed.
(e) Purified at point of use.
(f) Can be refrigerated to 40° F to compare with 3C1b, and dried to -80 F or lower to compare with 3C1c.
(g) Can be refrigerated to 40° F to compare with 3C2b, and dried to -80 F or lower to compare with 3C2c.
(h) Transportation costs are included.
(i) Costs are highly volume dependent. The volumes used for calculating price ranges are:

	Min.	Max.
Ammonia	Cylinder Lot	Tank Truck
Hydrogen - Bulk Compressed Gas	Cylinder Lot	500,000 CF/mo
Liquid	100,000 CF/mo	1,000,000 CF/mo
Nitrogen	400,000 CF/mo from Liquid N_2	25,000 CFH from On-Site N_2 Plant
Argon	Cylinder Lot	500,000 CF/mo
Helium	Cylinder Lot	3,000,000 CF/mo

(j) For carbonitriding, up to both 5% methane and 10% ammonia are added to the carrier gas.
(k) Dew points of -80 F or lower are achieved with molecular sieve driers having "closed-circuit reactivation," resulting in the approximate atmosphere-gas costs tabulated. Minus 100 F or better can be attained, however, by using driers with "bleed reactivation," in which dried gas is bled to waste through the bed being reactivated. This results in lower equipment costs but higher atmosphere-gas costs.

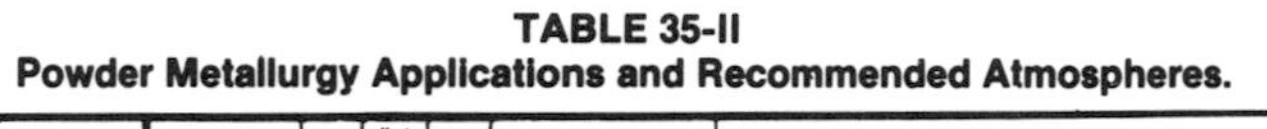

TABLE 35-II
Powder Metallurgy Applications and Recommended Atmospheres.

Atmosphere Gases (b) / Material	Annealing					Carburizing		Heat Treating		Infiltration	
	Copper	Iron—Carbonyl	Iron—Electrolytic	Steels—Low Carbon	Steels—Medium Carbon	Iron	Tungsten (with lamp black)	Steels-Carbon	Steels-Copper	Iron	Steels-Carbon
1. Hydrogen											
A. By Electrolysis of Water											
1. Direct from Cells											
a. Unpurified-Saturated	●	●	●	●						●	
b. Purified							●				
B. Bulk Compressed Gas	●	●	●	●			●			●	
C. By Catalytic Conversion of Hydrocarbons											
1. Unpurified—Saturated	●	●	●	●						●	
2. Dried							●				
D. From Liquid Hydrogen							●				
2. Nitrogen Based Enriched with											
A. Hydrogen or Dissociated NH_3	●	●	●	●			●		●	●	
1. With Methane Added					●			●			●
B. Endothermic Gas	●			●	●			●	●	●	●
C. Methanol	●			●	●			●	●	●	●
3. Ammonia Based											
A. Dissociated Ammonia											
1. As Reacted—Dry	●	●	●	●			●			●	
2. Moisture Added—Saturated											
B. Burned Dissociated Ammonia											
1. Rich—Saturated	●	●	●	●						●	
2. Lean-Saturated	●										
C. Direct Catalytic Conversion of Ammonia and Air											
1. Rich											
a. As Reacted—Saturated	●	●	●	●						●	
b. Cooled to 40° F											
c. Dried											
2. Lean											
a. As Reacted—Saturated	●										
b. Cooled to 40° F											
c. Dried											
4. Reformed Hydrocarbon Gases											
A. Exothermic Gas											
1. Rich											
a. As Reacted—Saturated	●			●						●	
b. Refrigerated											
2. Medium Rich—Saturated	●			●							
3. Lean—Saturated											
B. Purified Exothermic Gas											
1. Rich	●			●	●			●	●		●
2. Medium Rich											
a. As Reacted	●			●	●			●	●	●	●
b. Methane Added						●		●	●		
3. Lean	●										
C. Endothermic Gas											
1. Rich—Dry	●				●	●		●	●		●
2. Rich—Fairly Dry											
a. As Reacted	●				●			●	●	●	●
b. Methane Added						●					
3. Medium Rich—Saturated	●			●						●	
4. Lean-Saturated	●			●						●	
5. Argon—Liquid or Bulk Compressed Gas	●										
6. Helium—Bulk Compressed Gas	●										
7. Vacuum (Below 150 Microns)											
8. Air											
A. Normal											
B. Wet											

Atmosphere Gases (b) / Material	Reducing Oxides						
	Cobalt	Iron	Molybdenum	Nickel	Steels—Carbon & Alloy	Steels—Stainless	Tungsten
1. Hydrogen							
A. By Electrolysis of Water							
1. Direct from Cells							
a. Unpurified-Saturated		●		●			●
b. Purified	●		●			●	
B. Bulk Compressed Gas	●	●	●	●		●	●
C. By Catalytic Conversion of Hydrocarbons							
1. Unpurified—Saturated		●		●			●
2. Dried	●		●			●	
D. From Liquid Hydrogen	●		●			●	
2. Nitrogen Based Enriched with							
A. Hydrogen or Dissociated NH_3	●	●	●	●	●	●	●
1. With Methane Added					●		
B. Endothermic Gas		●		●	●		
C. Methanol		●		●	●		
3. Ammonia Based							
A. Dissociated Ammonia							
1. As Reacted—Dry	●	●	●	●		●	●
2. Moisture Added—Saturated							
B. Burned Dissociated Ammonia							
1. Rich—Saturated							
2. Lean-Saturated							
C. Direct Catalytic Conversion of Ammonia and Air							
1. Rich							
a. As Reacted—Saturated							
b. Cooled to 40° F							
c. Dried						●	
2. Lean							
a. As Reacted—Saturated							
b. Cooled to 40° F							
c. Dried							
4. Reformed Hydrocarbon Gases							
A. Exothermic Gas							
1. Rich							
a. As Reacted—Saturated							
b. Refrigerated							
2. Medium Rich—Saturated							
3. Lean—Saturated							
B. Purified Exothermic Gas							
1. Rich					●		
2. Medium Rich							
a. As Reacted					●		
b. Methane Added							
3. Lean							
C. Endothermic Gas							
1. Rich—Dry					●		
2. Rich—Fairly Dry							
a. As Reacted					●		
b. Methane Added							
3. Medium Rich—Saturated							
4. Lean-Saturated							
5. Argon—Liquid or Bulk Compressed Gas							
6. Helium—Bulk Compressed Gas							
7. Vacuum (Below 150 Microns)							
8. Air							
A. Normal							
B. Wet							

Atmosphere Gases (b) / Material	Sintering												
	Alnico	Beryllium	Brass	Bronze	Carbides of Refractory Metals	Copper	Iron	Iron-Copper	Aluminum	Metal-Ceramic Combinations (Cermets)	Molybdenum	Nickel	Niobium (Columbium)
1. Hydrogen													
A. By Electrolysis of Water													
1. Direct from Cells													
a. Unpurified-Saturated				●		●	●	●				●	
b. Purified	●		●		●				●		●		
B. Bulk Compressed Gas	●				●		●				●		
C. By Catalytic Conversion of Hydrocarbons													
1. Unpurified—Saturated				●		●	●	●				●	
2. Dried	●		●		●						●		
D. From Liquid Hydrogen	●		●		●				●		●		
2. Nitrogen Based Enriched with													
A. Hydrogen or Dissociated NH_3			●	●	●	●	●	●	●		●	●	
1. With Methane Added													
B. Endothermic Gas			●	●		●	●	●				●	
C. Methanol			●	●		●	●	●				●	
3. Ammonia Based													
A. Dissociated Ammonia													
1. As Reacted—Dry			●	●		●	●	●	●		●	●	
2. Moisture Added—Saturated													
B. Burned Dissociated Ammonia													
1. Rich—Saturated				●		●	●	●				●	
2. Lean-Saturated													
C. Direct Catalytic Conversion of Ammonia and Air													
1. Rich													
a. As Reacted—Saturated				●		●	●	●				●	
b. Cooled to 40° F													
c. Dried			●						●				
2. Lean													
a. As Reacted—Saturated													
b. Cooled to 40° F													
c. Dried													
4. Reformed Hydrocarbon Gases													
A. Exothermic Gas													
1. Rich													
a. As Reacted—Saturated				●		●	●	●				●	
b. Refrigerated													
2. Medium Rich—Saturated				●								●	
3. Lean—Saturated													
B. Purified Exothermic Gas													
1. Rich			●	●		●	●	●				●	
2. Medium Rich													
a. As Reacted			●	●		●	●	●				●	
b. Methane Added													
3. Lean							●						
C. Endothermic Gas													
1. Rich—Dry			●	●		●							
2. Rich—Fairly Dry													
a. As Reacted			●	●		●							
b. Methane Added													
3. Medium Rich—Saturated				●		●	●	●				●	
4. Lean-Saturated				●		●	●	●				●	
5. Argon—Liquid or Bulk Compressed Gas		●			●				●	●	●		●
6. Helium—Bulk Compressed Gas		●			●				●	●	●		●
7. Vacuum (Below 150 Microns)		●			●		●	●	●	●	●		●
8. Air													
A. Normal									●				
B. Wet													

Atmosphere Gases (b) / Material	Sintering (continued)											
	Silver	Steels—Carbon & Alloy	Steels—Stainless	Tantalum	Titanium	Thorium	Tungsten	Tungsten Alloys	Uranium	Uranium Oxide	Vanadium	Zirconium
1. Hydrogen												
A. By Electrolysis of Water												
1. Direct from Cells												
a. Unpurified-Saturated	●									●		
b. Purified			●				●	●	●			
B. Bulk Compressed Gas			●				●	●	●			
C. By Catalytic Conversion of Hydrocarbons												
1. Unpurified—Saturated	●									●		
2. Dried			●				●	●	●			
D. From Liquid Hydrogen			●				●	●	●			
2. Nitrogen Based Enriched with												
A. Hydrogen or Dissociated NH_3	●		●				●	●		●		
1. With Methane Added		●										
B. Endothermic Gas	●	●										
C. Methanol	●	●										
3. Ammonia Based												
A. Dissociated Ammonia												
1. As Reacted—Dry	●		●				●	●		●		
2. Moisture Added—Saturated												
B. Burned Dissociated Ammonia												
1. Rich—Saturated	●									●		
2. Lean-Saturated												
C. Direct Catalytic Conversion of Ammonia and Air												
1. Rich												
a. As Reacted—Saturated	●									●		
b. Cooled to 40° F												
c. Dried			●									
2. Lean												
a. As Reacted—Saturated												
b. Cooled to 40° F												
c. Dried												
4. Reformed Hydrocarbon Gases												
A. Exothermic Gas												
1. Rich												
a. As Reacted—Saturated	●											
b. Refrigerated												
2. Medium Rich—Saturated	●											
3. Lean—Saturated												
B. Purified Exothermic Gas												
1. Rich	●	●										
2. Medium Rich												
a. As Reacted	●	●										
b. Methane Added												
3. Lean	●											
C. Endothermic Gas												
1. Rich—Dry		●										
2. Rich—Fairly Dry												
a. As Reacted		●										
b. Methane Added												
3. Medium Rich—Saturated	●											
4. Lean-Saturated	●											
5. Argon—Liquid or Bulk Compressed Gas			●	●	●	●	●	●	●	●	●	●
6. Helium—Bulk Compressed Gas			●	●	●	●	●	●	●	●	●	●
7. Vacuum (Below 150 Microns)		●	●	●	●	●	●	●	●	●	●	●
8. Air												
A. Normal												
B. Wet												

(a) Where wet gases are recommended, dry gases will serve equally well, possibly better. Exceptions are in Section 4, wherein drier exothermic gases have higher carbon potential than wet gases, which may result in carburizing of low-carbon materials.
Applications of dry gases often require that the furnaces be built and operated in a manner that will contribute to holding high-purity atmospheres within the furnaces.
(b) For approximate compositions and costs of atmosphere gases see **Table 35-I**.

hydrogen or dissociated ammonia or nitrogen-hydrogen blends. Vacuum is not suitable for sintering brass because the zinc will volatilize.

Oxidation of compacts made of iron or iron-graphite mixes is caused by oxygen, water vapor, and carbon dioxide. If present in unsatisfactory proportions with respect to hydrogen and carbon monoxide, they can cause discoloration or scaling. Iron oxides are reduced by hydrogen, carbon monoxide and carbon. Decarburization is caused by oxygen, water vapor, and carbon dioxide. Carburization, on the other hand, is caused by carbon monoxide and hydrocarbons such as methane. Vacuum sintering will neither carburize nor decarburize iron-graphite mixes.

35.10 Reformed Hydrocarbon Gases

There is a family of low-cost gaseous mixtures made by reforming hydrocarbon gases which will be described in this section. Each type of gas is named from the kind of reaction by which it is produced, has wide usage in various types of powder metallurgy sintering, and has its own advantages and limitations. The gases are called "exothermic gas," "purified exothermic gas," and "endothermic gas."

35.11 Exothermic Gas

If air and fuel gas are mixed in proper proportions for complete combustion, and the mixture is burned in a refractory-lined reaction chamber, a water-jacket may be needed because so much heat is given off by the reaction. When heat is generated, it is termed an "exothermic reaction." Accordingly, for convenience in protective-atmosphere work, the products of combustion obtained in this manner are labeled as being "exothermic gas."

Figure 35-1 indicates, in the right half, the composition, with the exception of nitrogen, of exothermic gas from the producer under various operating conditions. (Endothermic gas, to the left, is discussed later). After adding the percentage of the various constituents shown on the chart for any single operating condition (air: gas ratio), the difference between 100 and this sum is the percentage of nitrogen. In all cases, nitrogen is actually the largest single constituent. Percentages of the constituents can be read in the vertical column at the left of the chart. The numbers across the bottom indicate the ratios of the input mixtures of air and natural gas, which is the most common fuel gas used. Each number indicates the volumes of air compared to one volume of 1000Btu (252 Kg Cal) natural gas in the mixture, such as 6:1 at the center of the chart or 10:1 at the right. Other charts are available to show similar compositions when using different hydrocarbon fuel gases, such as coke-oven, propane, or butane gas.

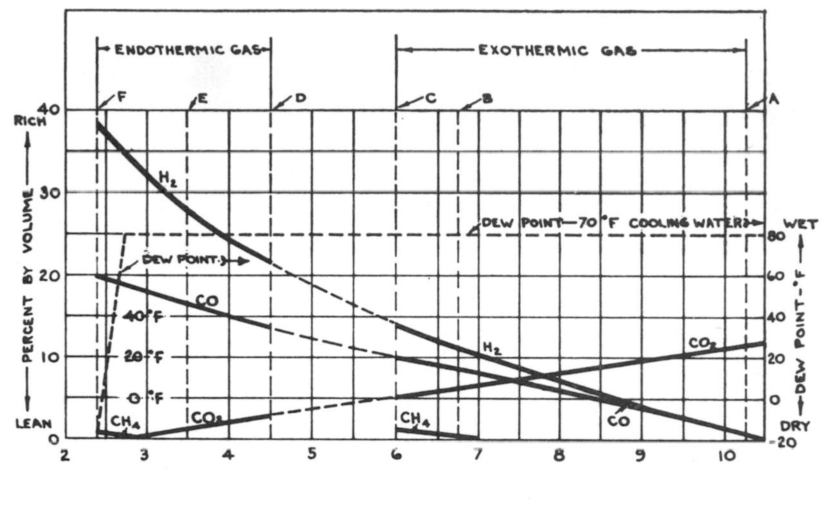

FIGURE 35-1

Constituents of endothermic and exothermic gases.

COMPLETE COMBUSTION
10 vol = 1 vol
(AIR) (GAS)

As indicated at the extreme right of figure 35-1 nearly complete combustion, or the point at which practically all combustibles in the fuel gas are burned up, takes place with a ratio of about 10.25 volume of air to 1 volume of natural gas. The composition of the resulting output gaseous mixture from the exothermic gas produced, as indicated by the points at which the verticle dash-line A intersects the curves, is about 0.7% hydrogen, 0.7% carbon monoxide, 11.5% carbon dioxide, and 87.1% nitrogen (not shown on the chart). This gas is saturated with water vapor at about 10F (-12C) above the temperature of the cooling water used in the surface cooler of the producer. In figure 35-1, the dew point is indicated to be 80F (27C) with cooling-water temperature of 70F (21C), which is low compared with the high amount (about 18%) of moisture initially formed in the reaction. In other words, the cooler condenses most of the moisture.

The complete combustion of natural gas, which is mostly methane, with air which is about 21% oxygen, 79% nitrogen, or 1 part oxygen to 3.8 parts nitrogen, is represented by the equation:

$$CH_4 + \text{air}\ (2O_2 + 7.6\ N_2) = CO_2 + 2\ H_2O + N_2 \qquad \textit{Eq. 35-1}$$

The gaseous mixture at vertical line A is relatively inert to some materials at elevated temperatures. For example, it will not oxidize hot copper, tin, or silver. It will, however, oxidize hot iron and the reactive metals because of the high percentages of carbon dioxide and water as opposed to the extremely low percentages of reducing components, hydrogen and carbon monoxide.

Being lean in fuel gas, this reformed mixture at A is called "lean exothermic gas." It is the lowest cost atmosphere gas available.

Actually, lean exothermic gas has very little application in powder metallurgy. Where it can be used, most individuals prefer to have a mixture richer in hydrogen to help overcome oxidizing effects of impurities which infiltrate into the furnace atmosphere.

When sintering bronze parts, for example, some users prefer to operate at about 6.5 or 7:1 air-gas ratio. As indicated in figure 35-1, using a mean ratio of 6.75:1, at vertical dash-line B, this "medium-rich" mixture has a composition of about 11.5% hydrogen, 8.8% carbon monoxide, 6.3% carbon dioxide, 0.3% methane, and 73.1% nitrogen. This gas is sometimes refrigerated, in the belief that reduction of the water content to about +40F (4.4C) dew point will give better results on the product being treated.

The richest exothermic gas is obtained by still further reducing the air to a 6:1 air-gas ratio, as indicated by the vertical dash-line C at the center in figure 35-1. Here the composition is about 14% hydrogen, 10% carbon monoxide, 1% methane, 5% carbon dioxide, and 70% nitrogen.

Carbon dioxide and water combine with and remove carbon from the surfaces of carbon-steel compacts being sintered. For example:

$$CO_2 + C \text{ (free graphite)} = 2\,CO \qquad \textit{Eq. 35-2}$$

$$CO_2 + Fe_3C = 3Fe + 2CO \qquad \textit{Eq. 35-3}$$

This removal of carbon from the surface results in the parts having reduced strength and a soft surface subject to rapid wear. When sintered and hardened without decarburization, parts are hard to the surface and have good wear-resistance and strength.

When an atmosphere gas neither carburizes nor decarburizes a ferrous metal, the gas is said to have a "carbon potential" equal to the carbon content of the metal. For example, an atmosphere neutral to compacts with 0.40% carbon at the temperature involved, would have a carbon potential of 0.40% carbon.

Sometimes exothermic gas is dried to help improve its properties, but drying alone generally is considered to be not too helpful, because water can be formed in the furnace by the "water-gas reaction" between hydrogen and carbon dioxide; the equation for this is:

$$H_2 + CO_2 = CO + H_2O \qquad \textit{Eq. 35-4}$$

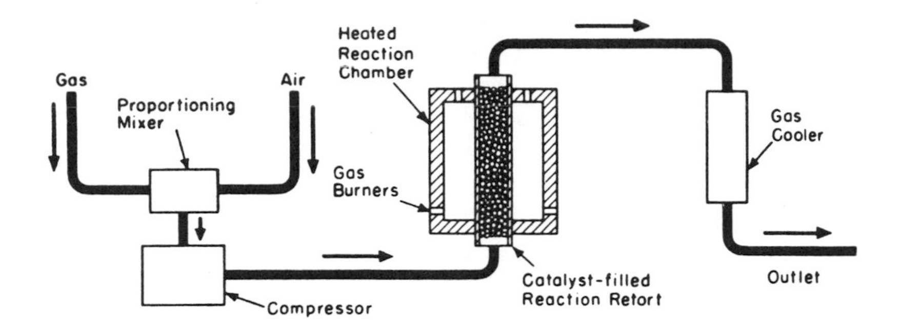

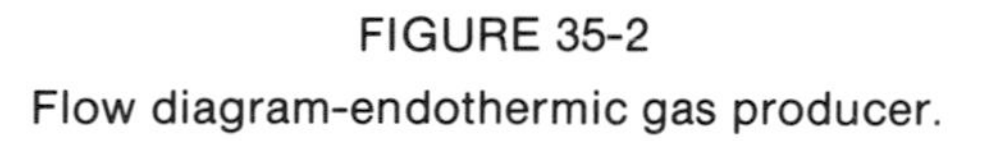
FIGURE 35-2

Flow diagram-endothermic gas producer.

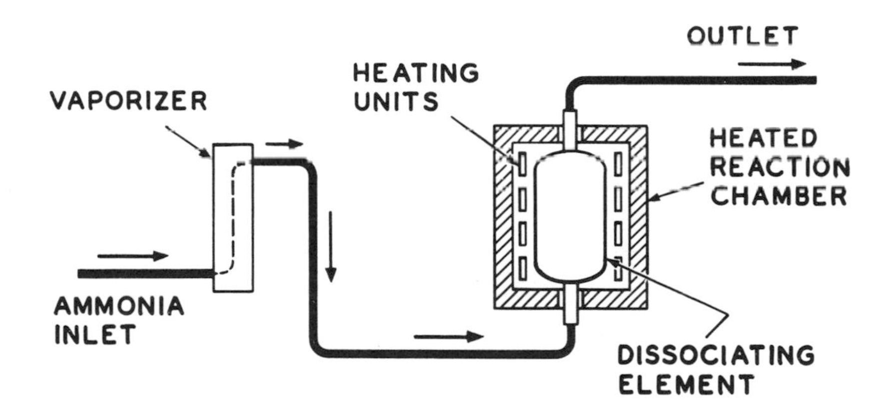

FIGURE 35-3

Ammonia dissociator.

Thus, to purify the gas effectively, carbon dioxide also should be removed to obtain a stable, dry furnace atmosphere.

A hydrocarbon fuel gas occasionally contains sulphur, which is commonly the case with coke-oven gas. Generally the sulphur removal is not necessary when sintering iron parts.

Benefits are low flammability of the gas, which contributes to safety in its use; low thermal conductivity, which helps the operating economy of the furnace with respect to thermal losses; and relatively trouble-free operation of the exothermic gas producer.

Rich exothermic gas is useful for sintering copper, bronze, silver, iron-copper and iron. Its carbon dioxide and water content, however, make it unsuitable for some P/M parts. Brass may lose zinc and become discolored and have inferior mechanical properties, and carbon steel may be decarburized. Steel parts may be recarburized, however, before or during heat treatment.

Figure 35-4 shows a simplified flow diagram of a typical exothermic gas producer. At the left, gas and air enter through flowmeters (not shown) and then into a proportioning mixer or carburetor. A compressor boosts the pressure of the mixture to force it through a fire-check and burner at the entrance to the reaction chamber. It then passes through the catalyst bed, which is heated by the products of combustion to help the reactions to completion in a small space. The products of combustion then pass through a gas cooler, to condense most of the water then on to the furnace, unless some additional form of purification is necessary.

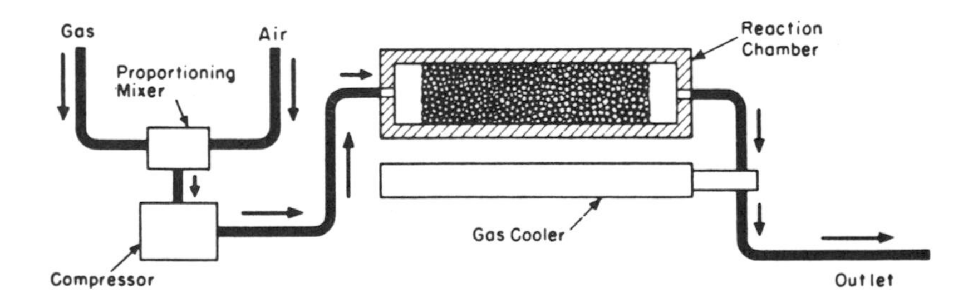

FIGURE 35-4

Flow diagram-exothermic gas producer.

35.12 Purified Exothermic Gas

It is sometimes desirable to remove carbon dioxide and water from rich exothermic gas for sintering applications. Gas producers with accessories are used in such cases to make purified rich exothermic gas, free from carbon dioxide and water. In typical instances, the input ratio of air to natural gas is 6.5 or 7:1. Taking an average of 6.75:1, the composition is about 12.0% hydrogen, 9.0% carbon monoxide, 0.2% methane, and 78.8% nitrogen. The dew point usually is -40F (-40C) or lower.

Carburizing of carbon-steel parts results in growth and possible blistering; decarburizing causes shrinkage—either condition results in wide dimensional changes. Purified rich exothermic gas, being only mildly reactive and quite stable, creates no noticeable carburizing or decarburizing of iron or carbon-steel parts, contributing to relatively easy dimensional control. It is assumed that there might be slight carburizing of iron parts or decarburizing of high-carbon iron parts, but is so nominal that it cannot be detected in photomicrographs, physical properties or performance characteristics. Accordingly, for practical purposes, the gas is assumed to be neutral over the entire range from no carbon to high carbon when sintering P/M parts. No adjustment, change, or control of the purified rich exothermic gas is necessary, other than to keep it carbon dioxide-free and dry, which conditions are assured by means of the auxillary purifying equipment.

Of course, to obtain reliable results, uniform conditions must be held within the furnace. In a tight furnace, with normally-closed end doors, such as the box- and roller-hearth types, dew points of -20 to +25 F (-29 to -4 C) and 0.0 to 0.2% carbon dioxide are commonly maintained in production. Open-ended mesh-belt type furnaces require substantially higher flow rates of atmosphere gas and better control of room drafts, to achieve comparable results.

When sintering carbon-steel compacts, regardless of the type of atmosphere gas used, generally there is a loss in the amount of graphite from the original mix to the final amount of combined carbon. This is attributed normally to reactions of the graphite with oxygen in various forms within the compact, to create carbon dioxide. The extent of graphite loss depends upon such factors as the amount of air, moisture, and oxides, the degree of activity of the grades of graphite and iron used, temperature, and time. It is common to start with about 1¼ to 2 times as much graphite in the green compacts as desired in combined carbon in the final sintered parts. When using purified rich exothermic gas, the final combined carbon is uniformly dispersed from center to surface of the parts.

Variations in composition of fuel gas supplied to the purified exothermic gas producer naturally result in variations in the output gas mixture of hydrogen, carbon monoxide, and nitrogen. There are, however, no variations in the carbon dioxide and water, since these objectionable constituents are consistently removed by the purifying system. Accordingly, variations in the input fuel-gas supply create no noticeable effect on the product or in the operation of the equipment. The fact that purified exothermic gas is always carbon dioxide-free and dry is, therefore, an important determining factor in its selection for critical applications. Also, this atmosphere gas has low flammability and thermal conductivity, which are desirable attributes.

In purified exothermic gas producers, sooting of the catalyst and consequent deterioration of the effluent product gas, requiring frequent "burning out" of the reaction chambers for restoration, is rare. Such producers commonly run months or even years continuously without the need for a "burning-out" procedure.

Some users find it desirable to "burn-out" the furnaces briefly each weekend or so, when using dry non-decarburizing atmospheres. This is done to remove accumulations of carbon, caused by distillation of lubricants from the compacts, even though substantial preheaters or burnoff chambers are used ahead of the high-heat chambers. Burning out is accomplished by cooling the furnace to about 1300 F (700 C), opening the end doors, lighting the gas flowing out the ends, and then shutting off the gas supply. As the gas burns in, air enters the heating chamber and comes in contact with incandescent carbon deposits. This causes a temporary increase in temperature and cleans out the carbon more effectively than can be achieved by any other method.

Purified rich exothermic gas has a wide range of applications. Its most common uses are for sintering iron, iron-copper, carbon-steel, copper-steel, and for copper infiltrating iron and carbon-steel parts. It is also used for sintering bronze compacts, and in a few instances, brass compacts on trays under covers.

A simplified flow diagram for a purified exothermic gas producer is shown in figure 35-5. The initial stages are similar to those for the exothermic gas producer. Gas and air are metered, and pass through a proportioning mixture or carburetor, a compressor, a fire-check, and a burner, then into the reaction chamber. In this case, the reaction chamber is built into a boiler, so that the heat from the exothermic gas is given off to the carbon dioxide-laden chemical solution in order to boil the solution and help drive off the carbon dioxide. This is an economy measure as well, by utilizing heat that would otherwise be wasted.

The products of combustion pass on through a gas cooler and up through a carbon dioxide absorbing tower, through which is sprayed the carbon dioxide-free solution taken from the bottom of the reactivating tower. At this point, the gas leaves the producer and goes to drying equipment,which commonly consists of a refrigeration-type gas cooler to reduce the dew point

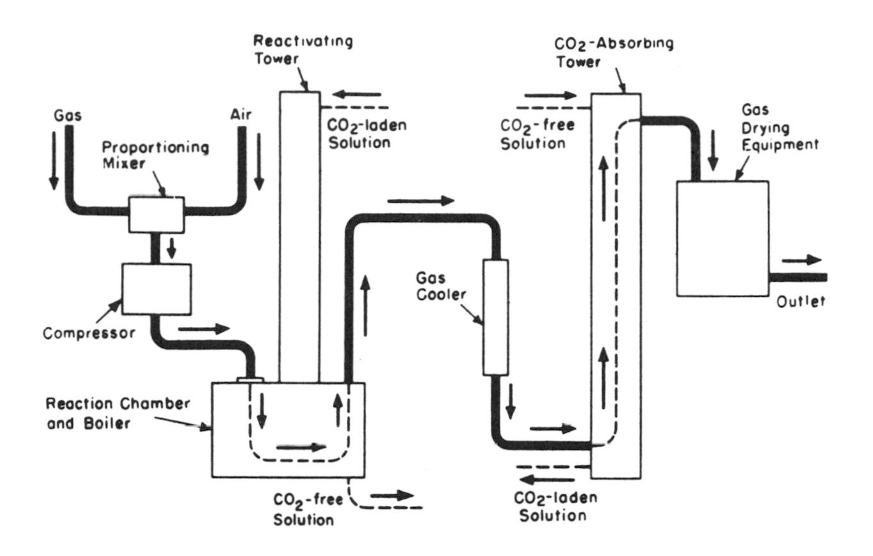

FIGURE 35-5
Flow diagram-purified exothermic gas producer.

to 40 F (4.4 C), and an activated-alumina dryer to further drop it to -40 F (-40 C) or lower. By thus scrubbing the undesirable constituents from the gas, the gas is always carbon dioxide-free and dry.

The relatively cool, carbon dioxide-laden solution from the bottom of the absorbing tower is passed through a heat exchanger, where it is heated by the boiling hot solution leaving the bottom of the reactivating tower. The hot carbon dioxide-laden solution is then charged into the top of the reactivating tower, is sprayed down over refractories where it gives up most of its carbon dioxide, and finally it runs down into the boiler. The solution, which is 10 to 15% monoethanolamine (MEA) in water, is constantly recirculated, and the operation is continuous and automatic.

Although the purified exothermic gas producer is rather complex at first glance, operators become acquainted with its flow circuits and accessories quickly, and find it to be a practical and dependable machine to manage. Routine checks of temperature and flows usually give adequate assurance of proper operation. About every eight hours the dual-chamber drier is switched over, a fresh drying tower is cut into the circuit, and the moisture-loaded tower is heated to drive out accumulated moisture. Simple weekly tests are made for determining the effectiveness of the solution.

35.13 Endothermic Gas

In the operation of an exothermic gas producer, the minimum ratio of air to natural gas is about 6:1. At lower ratios the heat given off by the combustion reaction is insufficient to sustain the reaction. However, if an additional source of heat is used to supply the needs of the reaction and provide heat for the thermal losses of the chamber, the air-gas ratio can be reduced to any value desired. Such a reaction, which requires the addition of heat for its completion, is called "endothermic." Thus the products of reacting a hydrocarbon gas and air over a catalyst with heat added, are commonly known as "endothermic gas."

In figure 35-1, the compositions of endothermic gas at various input air-gas ratios are shown. The maximum ratio at which some endothermic gas producers operate is 4.5:1 on vertical dash-line D. These units have silicon-carbide retorts. Vertical dash-line F is about the minimum ratio, 2.4:1, at which endothermic gas producers will function.

At the maximum ratio of 4.5:1, the composition of the endothermic gas is about 21.5% hydrogen, 13.8% carbon monoxide, 3% carbon dioxide, and 61.7% nitrogen. This gas is saturated with water at about 10 F (-12 C) above the temperature of the cooling water used in the surface cooler of the producer. Figure 35-1 shows 80 F (27 C) dew point with 70 F (21 C) water, at high ratios. At 2.4:1 ratio, the composition is about 38% hydrogen, 20% carbon monoxide, 0.5% methane, 45.1% nitrogen. Due to a characteristic of the reaction, this gas is very dry, having a dew point of 10 to -10 F (-12 to -23 C), without the need for auxiliary drying equipment. This dry, high-carbon monoxide mixture has important and desirable properties. The particular thing to note at the moment is that at 4:5:1 and down to 2.75:1 ratio, the gas is wet, while at 2.4:1 it is dry.

The overall chemical reaction between natural gas (methane) and air, taking place in the endothermic gas producer, is represented by the equation:

$$2CH_4 + \text{air}\ (0_2 + 3.8N_2) = 2CO + 4H_2 + 3.8N_2 \qquad \textit{Eq. 35-5}$$

By using an adequate catalyst and a sufficiently high reaction temperature with an accurately held air-gas ratio, the composition of the product gas can be controlled within desired limits. It will be noted in figure 35-1 that from 2.4:1 to about 2.9:1 ratios, the carbon dioxide content is approximately zero, but from 2.9:1 to 4.5:1 the carbon dioxide gradually creeps up to about 3%. Also, at 2.4:1 there is about 0.5% methane, but at about 2.9:1 the methane decreases to zero.

In figure 35-1, it is seen also that the dew point drops off rapidly between the air-gas ratios of about 2.75:1, at which it is about 80 F (27 C), depending upon the cooling-water temperature and 2.4:1, at which it is about -10 F (-23 C). This range is useful for making gases with correct carbon potential. It is seen by the dotted dew point curve that minor changes in the ratio result in major changes in dew point. Also, it is known that the lower the water (and carbon dioxide), the higher the carbon potential, and vice versa. If the carbon potential of the gas is not in equilibrium with the carbon content of the metal, then the gas may either decarburize or carburize the work. The porosity of most P/M parts makes them more sensitive to changes in carbon potential than solid steels, and undesirable effects penetrate more deeply.

Changes in composition of the gas supplied to the generator can have the same over-all effect as changing the air-gas ratio. For best results it is necessary to hold a constant ratio and to use a fuel gas of constant composition. Gas producers are available which hold gas ratios after being set. Some natural gas supplies are constant, but others are subject to unannounced variations in composition. Best assurance of constant fuel-gas composition is sometimes obtained by resorting to use of propane made from natural gas, which is quite uniform in composition.

For critical jobs, dew points or carbon dioxide content of the atmosphere gas should be checked frequently. Any drift from normal can be corrected before trouble develops. Also automatic carbon-potential controllers, which regulate either dew point or carbon dioxide, are useful.

In sintering furnace applications, care must be taken to avoid contamination of the furnace-atmosphere samples by lubricants distilled out of the work. Typical distillation products are carbon and zinc. Such contaminants can upset the accuracy of the dew point controller. The critical zone of control is the high-heat chamber. Therefore an adequate auxiliary pre-heat or burn-off chamber is necessary to remove the lubricants before the work reaches the high-heat chamber. Also, ample flow of atmosphere gas through the high-heat chamber is needed to sweep any possible contaminants out through the end where the work enters.

It is apparent from figure 35-1 that wet endothermic gas at 3.5:1 ratio, vertical dash-line E, which is about the highest ratio used with alloy retorts in producers, is fairly high in reducing properties; it contains about 27.6% hydrogen and 16.5% carbon monoxide, totalling about 44.1% reducing constituents. This saturated gas, also containing a small amount of carbon dioxide, has no tendency to carburize low-carbon iron parts. Its highly-reducing property makes it quite useful in general P/M applications.

Because of its rather high hydrogen and carbon monoxide content, endothermic gas is quite flammable and must be handled with the same respect as hydrogen and dissociated ammonia. Dry endothermic gas is somewhat lighter than air, with a specific gravity of 0.622. Its thermal conductivity with respect to air is about 3.23:1.

Applications of wet endothermic gas which has low carbon potential, include sintering compacts made of iron, iron-copper, copper, bronze, silver, nickel, etc., and infiltrating iron parts with molten copper. Uses of dry endothermic gas which can have medium or high carbon potential, include annealing of low-alloy steel powders with medium carbon; sintering compacts of carbon-steel, copper-steel, copper, bronze, brass, etc.; and infiltrating carbon-steel components. It is quite versatile and has a wide field of usefulness.

A simplified diagram of an endothermic gas producer is shown in figure 35-2. As with other producers, hydrocarbon fuel gas and air are measured, mixed, and compressed, then passed into a reaction chamber. In this case, the catalyst is retained in a retort of silicon carbide or heat-resisting alloy. The retort is surrounded by the reaction chamber, heated with either gas burners or electric heating units, in the temperature range of about 1800-2200 F (980 to 1200 C), depending upon the equipment and requirements. Following the reaction in the retort, the gas is cooled and goes on to the point of use.

35.20 Hydrogen-Nitrogen Mixtures

Mixtures of hydrogen and nitrogen can be obtained from ammonia at lower cost than cylinder hydrogen. In many instances these mixtures serve equally as well as hydrogen, but sometimes they have unfavorable metallurgical reactions in nitriding the products. Therefore, each application is generally considered on its own merits.

35.21 Dissociated Ammonia:

Dissociated ammonia consists of 75% hydrogen-25% nitrogen. Normally, the dissociation or "cracking" of ammonia with heat, over a catalyst, is 99.95% or more complete. Thus there is only a trace (0.05% or less) of ammonia in the gas; oxygen is 0.0% and the moisture content is indicated by dew points of -40 to -60F (-40 to -51C).

Should the small amount of raw ammonia in the gas be objectionable, all traces can be eliminated by passing the gas through either: water, which will absorb ammonia but increase the moisture content (which can be removed); or activated alumina, or a molecular sieve, which will absorb ammonia, and also water, thus further drying the gas.

Because of its high hydrogen content, dissociated ammonia at elevated temperatures is highly flammable. Its specific gravity is 0.295 and thermal conductivity 5.507 compared to air, which is 1.0. Ammonia is generally available in cylinders containing 150 pounds (68 kg), in tank-truck deliveries of 2000 to 4000 pounds (907 to 1814 kg), or in bulk tank-truck or railway tank-car deliveries into the purchaser's storage facilities, in lots of 26,000 and 52,000 pounds (12,000 to 24,000 kg).

The uses of dissociated ammonia overlap, to a great extent, those of hydrogen and some of the low-cost gaseous mixtures. Typical appliations of dissociated ammonia include: sintering of brass, copper, iron, iron-copper, tungsten and tungsten alloys, aluminum and aluminum alloys, and stainless steels. Dissociated ammonia is sometimes avoided where molecular nitrogen causes nitriding and consequent hardness and embrittlement as in sintering stainless steel or molybdenum compacts.

A simplified flow diagram of an ammonia dissociator is shown in figure 35-3. Liquid ammonia from the tank enters a vaporizer at high pressure, where heat converts the liquid into vapor. The pressure of the vapor is then reduced in an expansion valve and the low-pressure vapor passes through a dissociator element filled with catalyst. Here, within the heated reaction chamber held at 1650 to 1850F (900 to 1010C), the ammonia is dissociated into its constituents hydrogen and nitrogen. The gaseous mixture then passes from the outlet of the dissociating element to the furnace.

35.22 Burned Dissociated Ammonia:

Burning of dissociated ammonia pre-mixed with air, in a reaction chamber reduces the hydrogen level to 24% at the upper limit down to 0.5% at the lower limit, depending upon the amount of air. The mixture is saturated with water, but it can be dried if necessary.

These low hydrogen-nitrogen mixtures are less highly reducing and less flammable than those with higher hydrogen. In fact, they are non-flammable when the hydrogen is at or near the lower limit.

The equipment in which dissociated ammonia is burned is basically the same as the exothermic gas producer shown in figure 35.4, and described in that related text. In this instance, dissociated ammonia takes the place of the hydrocarbon fuel gas entering the system.

35.23 Direct Catalytic Conversion of Ammonia and Air

Similar mixtures varying from 25% to 0.5% hydrogen, remainder nitrogen, are made by direct catalytic oxidation of ammonia with air, as compared to the two step process of dissociation plus combustion as just described.

Direct catalytic generators are available with various modifications. In a typical unit, ammonia vapor and air are mixed and fed into a catalytic chamber provided with a starting heater to hold 250F (120C) at the outset. Once the reaction begins it gives off heat and is self-supporting. The hydrogen-nitrogen-ammonia mixture from this first stage converter is combined with additional air and passed into a second stage converter in which the reaction goes to completion. The mixed hydrogen-nitrogen gas then goes to the furnace. Some units use ammonia vapor, in which case the output gas is saturated with water. Some use liquid ammonia, the expansion of which, in a vaporizer, refrigerates the output gas to a dew point of about 40F (4.4C) and requires a smaller drier if dry gas is needed.

35.30 Nitrogen Based

Nitrogen is mixed with enriching gases for sintering a variety of ferrous and nonferrous materials. Because it is nonflammable, nitrogen is also used as a safety purge for flammable atmospheres. Nitrogen is produced through the cryogenic separation of air. It is usually stored and transported as a liquid. Liquid storage vessels are the most economical source for most sintering facilities. For larger requirements, on-site nitrogen plants provide gaseous nitrogen at a lower cost. The storage vessel or on-site plants are most commonly owned and maintained by the nitrogen supplier. No gas producers are required for nitrogen-based atmospheres unless a generated gas is used for enrichment.

Liquid nitrogen contains less than .001% oxygen and has a dew point of less than -90° F (-68C). Nitrogen alone will not reduce, oxidize, carburize, or decarburize. Enriching gases must be added to supply these properties and to counter the detrimental effects of oxygen or water vapor which enter the furnace. Nitrogen-based atmospheres rely on the high purity of nitrogen to facilitate reduction and to help prevent oxidation or decarburization. By utilizing a low dew point, less hydrogen is required in the furnace to maintain a proper reducing potential as measured by the H_2/H_2O ratio. Similarly, the low levels of water vapor and CO_2 in the atmosphere allow proper carbon control with only small additions of carbon monoxide or gases containing hydrocarbons. The sintering atmosphere generally consists of 75% to 95% nitrogen to take advantage of its purity, although the composition may be adjusted as desired. The amount of nitrogen and enriching gas is controlled with a blend panel located at the furnace.

The type and amount of enriching gas is determined by the reducing potential and carbon potential required. The most common types of enriching additives are hydrogen, dissociated ammonia, endothermic gas, and methanol. Hydrogen is supplied as a gas or liquid and is stored outside the plant. Methanol is supplied as a liquid which vaporizes and dissociates in the hot zone of the furnace to form carbon monoxide, hydrogen, and trace amounts of water and carbon dioxide.

Nitrogen-hydrogen or nitrogen-dissociated ammonia blends provide a low dew point reducing atmosphere with neutral carbon potential. Small additions of natural gas or other hydrocarbons are added when a higher carbon potential is required. Nitrogen-endothermic and nitrogen-methanol blends provide a reducing atmosphere with a medium carbon potential. Small hydrocarbon additions are made also to these atmospheres if an increased carbon potential is desired. During idle periods, the enriching gas may be turned off and a reduced flow of nitrogen is used to protect the furnace and to facilitate a faster start-up.

With the proper choice of enrichment gas, nitrogen-based atmospheres can be used to sinter and infiltrate iron, carbon steel, and other ferrous and nonferrous alloys. Stainless steels and refractory metals can be sintered in nitrogen-based atmospheres when nitriding is not critical.

35.40 Hydrogen

At elevated temperatures hydrogen is highly reducing to oxides of some metals, such as iron and copper. This property, combined with the ready availability of hydrogen has been largely responsible for its wide use.

Hydrogen is very flammable, having an extremely high rate of flame propagation. Although this explosive property requires that it be handled with respect, it has been used in large production furnaces for many years with a good safety record. Because of its high rate of flame propagation, hydrogen burns with a short, hot flame immediately upon contact with air. The flame is an almost colorless blue.

This gas is very light in weight—in fact, it is the lightest element known. Its specific gravity is only 0.069 as compared to 1.0 for air. It is easily displaced by air, and rushes out the top of the furnace door openings rapidly when free to do so.

Hydrogen is an excellent heat conductor. The thermal conductivity is seven times that of air. Hence it accelerates both the heating and cooling rates of the work in furnaces. The thermal losses in furnaces are higher with hydrogen atmospheres than when using heavier, less conductive gases.

Typical applications of hydrogen, sometimes unpurified, sometimes purified, in the manufacture of powders, are in the reduction of oxides of iron, molybdenum, tungsten, cobalt, nickel and 18-8 chromium-nickel stainless steel, annealing of electrolytic- and carbonyl-iron powders, and carburizing of tungsten powders in lamp black to form tungsten carbide. Hydrogen is used for sintering compacts of molybdenum, tungsten, tungsten carbides, stainless steels, brass, aluminum and aluminum alloys. Hydrogen is blended with cryogenically produced nitrogen for sintering iron, steel, brass, bronze, copper and stainless steel.

Hydrogen provided as a cryogenic liquid or in bulk gaseous trailers is the most economical source for requirements up to 8 to 10 million ft^3 (225 to 290

$\times$ 10^3 m^3). The storage vessels are most commonly owned and maintained by the hydrogen supplier. Cryogenically produced hydrogen is transported and stored as a liquid. Liquid hydrogen contains less than .002% impurities, with an oxygen content of less than .0002% and a dew point of less than -90F (-68C). Bulk (compressed gas) hydrogen contains less than .05% impurities with an oxygen content of less than .0005% and a dew point of less than -90F (-68C).

For larger requirements, hydrogen is produced on-site by electrolysis or by the steam reforming of natural gas. In electrolysis, electric current passes through electrolyte in cells, causing hydrogen to collect at one electrode in each cell, and oxygen at the other. The electrolyte is either sodium hydroxide or potassium hydroxide in distilled water.

Production by steam reforming natural gas involves heating a steam-natural gas mixture and passing it through a catalyst to produce hydrogen plus carbon monoxide and a small amount of carbon dioxide. This mixture with steam is further catalyzed to convert carbon monoxide to carbon dioxide, accompanied by production of an additional volume of hydrogen equal to that of the carbon monoxide. The resulting mixture of hydrogen plus carbon dioxide is passed through a chemical absorbing tower which removes the carbon dioxide, leaving hydrogen and a slight amount of carbon monoxide as an impurity.

Hydrogen generally contains fractional percentages of impurities in the order of 0.1 to 0.5% which may or may not be troublesome. Impurities of o ygen and water vapor can be removed readily if troublesome. Impurities of carbon monoxide and methane result from some methods of manufacture, but attempts are seldom made to remove them.

The most accepted method at present to remove oxygen is to pass the gas through a palladium catalyst purifier, which forms water. The unit operates at room temperature and is very effective. It lowers the oxygen to less than one part per million. Hydrogen delivered from this purifier contains water vapor, of course; and if dry gas is necessary, moisture removal is required.

Hydrogen as manufactured generally is saturated with water vapor. The generators are very expensive, and are found only in plants having high consumption.

Moisture removal is customarily achieved by refrigeration or by adsorption depending upon the moisture content and volume of gas used. Line hydrogen from an in-plant producer, as contrasted to bottled gas, commonly is saturated with water at room temperature and at atmospheric pressure. At high rates of flow, say 1000 ft^3/hr (28 m^3/hr) or more, the gas first can be refrigerated to +40F (4C). This will remove about two-thirds of the water vapor.

For eliminating the remaining moisture in the two step procedure, or for drying all the way in one step, regenerative adsorption-type driers are used. Such driers usually contain beds of desiccant materials in one or two towers, depending on whether service is to be intermittent or continuous. A common desiccant is activated alumina, which will dry the gas very effectively. For critical applications requiring the driest possible gas, a material called "molecular sieve" is highly successful.

After a tower has adsorbed its full capacity of water in the capillary pores of the desiccant, the bed is heated and air or gas is circulated through it to reactivate the desiccant. The bed is then cooled, and is ready for use again.

35.50 Vacuum

Up to this point, various atmosphere gases used for protecting different materials while being heated and cooled have been discussed. Some materials, however, are highly reactive and will combine with some of these gases or their impurities even though the gases are "purified". Such reactions may seriously impair some properties of the materials. In many such cases, however, it has been found that these materials can be treated in a vacuum with considerable success.

Vacuum furnaces produce an environment which retains the proper chemistry of the parts during sintering. With vacuum, reduction of oxides takes place while the parts are at temperature, and outgassing of the parts also is accomplished. In vacuum sintering, the environment is such that it is neither carburizing nor decarburizing during the sintering process.

Vacuum furnace heating elements are usually made of graphite or refractory metals such as tungsten or molybdenum. For temperatures under 800° C nichrome elements are also used. Insulation may be fibrous graphite or ceramic or comprised of layers of radiation shields of stainless steel, molybdenum or tungsten.

Vacuum equipment is so flexible it can be adapted to the sintering and heat treating of alloyed components, e.g., first sintering, then precipitation-hardening, solution annealing (including oil quenching), and aging. All of this can be accomplished without unloading the furnace until the sequence is finished.

Considerable production and experimental sintering is being done in vacuum. Most of this work is on the reactive metals which are highly susceptible to the formation of hydrides, nitrides, or oxides in gaseous atmospheres. Good results are obtained by vacuum sintering refractory metal carbides, stainless steels, beryllium, titanium, zirconium, tantalum, niobium (columbium), vanadium, thorium, uranium, and metal-ceramic (cermet) combinations.

Protective atmospheres in furnaces normally are used at atmospheric pressure. Most high-vacuum work is done at pressures considerably lower than 1 mm Hg (133 Pa). Vacuum pressures for sintering often are expressed in terms of microns of Hg. One micron is 0.001 mm (0.133 Pa). (1 Torr = 1 mm = 1000 microns = 133.3 Pa)

To create a vacuum in a closed chamber air molecules must be removed. The more removed, the higher the vacuum. Normal atmospheric pressure is 101.3×10^3 or 29.9 in (760 mm) of mercury. One millimeter of mercury, then, is 1/760th of an atmosphere. This basic unit of vacuum measurement is called a Torr (133Pa). One micron of mercury is 1/1000 of a Torr (1 mm Hg). For example, 1 Torr = 1000 microns Hg or 1×10^3 Hg and 1 micron Hg = 1×10^{-3} Torr (0.133Pa). Note that when working with exponents of 10, a vacuum of 1×10^{-3} Torr represents a pressure level ten times lower (higher vacuum) than a vacuum of 1×10^{-2} Torr.

At atmospheric pressure 1 cc of air contains 3×10^{19} molecules moving around at random which results in a large number of molecular collisions. The mean free path (M.F.P.) between molecules is 2.6×10^{-6} in (0.066 mm). If this 1 cc container is evacuated to one micron (1×10^{-3} Torr or 0.133 Pa) which is a very good vacuum for most heat treating, there are still 4×10^{13} (over half) of the initial air molecules remaining. At this lower pressure, there is a tremendous increase in mean free path and the probability of molecular collision with the surface of a work piece is greatly reduced.

If the furnace is further pumped down to 10^{-9} Torr (133×10^{-9} Pa) which is one millionth of a micron, the number of molecules per cc is decreased to 4×10^{-7} and the mean free path is 5,000,000 cm or 30 miles. The path of the molecules is now limited by the walls of the chamber and not by collision with each other. Flow as we know it does not exist. This is the reason for the relatively large piping used in diffusion pump suction lines. The only molecules that a diffusion pump moves out of a vessel are those which drift into the pump suction and are captured. If the pump opening was not there, these molecules would rebound from the wall to other parts of the chamber.

For purposes of comparison, space vacuums are 10^{-8} microns at 200 miles (340 km) and 10^{-10} microns at 400 miles (645 km). Outer space is reported to be 10^{-16} microns.

A vacuum furnace at one micron (0.133 Pa) corresponds to an atmospheric dew point of -115F (-82C).

The vacuum pumping system is connected directly to the chamber in which the work is heated and cooled. The chamber usually is pumped down to the desired pressure before heat is applied. During heating, however, there may be bursts of absorbed gases from the work. These will increase the chamber pressure and may require that the temperature be held constant until the pressure is pumped down again to the desired level. Organic materials in pressed powder compacts are particularly prone to violent outgassing due to decomposition. It is not uncommon to use a moderate temperature presintering treatment in a pure gas atmosphere to drive off the volatiles thus avoiding overloading and/or contaminating the vacuum system. Even then, when the compacts are subsequently sintered in vacuum, it may be necessary to install a "cold trap" ahead of the vacuum pump to collect condensable vapors, or to use one of several other expedients to remove them. This will assure best pumping efficiency and minimize contamination of the vacuum-pump oil.

When metallic additives evolve they cause pumping system problems whereas organic residuals do not have this effect and vacuum pumps can accommodate their removal from the furnace environment without any problems. Metallic additives can also have a deleterious effect upon thermocouples and heating elements; hence, it is advisable that only organic-based additives be used in vacuum sintering operations.

Vacuum is often more economical than atmosphere gases, particularly bottled gas. The only operating costs involved in producing the vacuum are for electrical energy and oil for the pumps. Vacuum furnaces, however, are often more expensive than equivalent protective-atmosphere furnaces for the same work, due to the necessity for vacuum-tight welding, leak testing, and strengthening retorts and bases against high collapsing pressures.

Vacuum pumps commonly used are mechanical pumps and oil-vapor pumps. Two pumps of similar or unlike types are used often in tandem for powder metallurgy work, one (the mechanical "forepump"), having a dual purpose as "roughing and backing pump" to reduce the pressure to a point at which the vapor stream or diffusion pump can be cut in to further reduce the pressure.

Rotary, positive-displacement, mechanical pumps generally fall into three subdivisions: (1) single stage oil-sealed pumps for use from 760 Torr (101 $\times 10^3$ Pa) to 10 micron Hg McLeod (1.33 Pa), (2) compound oil-sealed pumps for use from 760 Torr to 0.2 microns Hg McLeod (0.026 Pa), and (3) oil free mechanical booster lobe pumps for use from approximately 50 Torr (6650 Pa) to below 0.1 micron Hg McLeod (0.0133 Pa). This last group (3) must be operated in tandem with a standard rotary oil-sealed pump, as it can not discharge to atmospheric pressure except in very small sizes.

Diffusion-ejector pumps and diffusion pumps with mechanical backing pumps handle large throughputs between the range of about 650 and 2 microns (86.5 to 0.266 Pa), and 10 to 0.1 microns Hg (1.33 to 0.0133 Pa), respectively. Oil diffusion pumps with mechanical backing pumps are effective from 100 microns down to 0.0001 microns (13.3 to 1.33×10^{-5} Pa). The actual pressure ranges vary somewhat with size and make. The ranges given here are typical.

Cryogenic pumps using surfaces refrigerated to liquid hydrogen temperatures (20° K) or below are also employed. The low temperature surface traps gas molecules by condensation.

35.60 Argon and Helium:

Argon (A) and helium (He) are nonflammable and are inert for almost all applications. They are used for sintering refractory and reactive metals (Table 35-II), and typically as a backfill in vacuum furnaces. Their cost generally prohibits use in continuous furnaces. Argon and helium are also used for inert gas atomization of reactive metals.

Argon is cryogenically produced from air. Its purity is very high, with less than .003% impurities including less than .0005% oxygen and a dew point lower than -90F (-68C). Its specific gravity is 1.379 g/cm^3. Its thermal conductivity, 0.745, is lower than that of air. Bulk argon is available as a compressed gas in cylinders or trailers. For larger volumes, liquid argon is more economical.

36.10 INSTRUMENTS FOR ANALYZING AND CONTROLLING GASES

There is a wide variety of instruments for analyzing and controlling protective atmosphere gases for powder metallurgy sintering. A few typical ones will be discussed to illustrate the convenience and helpfulness of having such equipment available, and to describe the general principles on which they operate.

36.11 Orsat-Type Analyzers

Complete analysis of a gaseous mixture generally is divided into the absorption phase and the explosion or burning phase. Percentages of carbon dioxide, oxygen, and carbon monoxide are first determined by chemical absorption in the order named. This is generally called the Orsat principle. Orsat analyzers can be obtained for measuring carbon dioxide, oxygen and carbon monoxide, or for only carbon dioxide or oxygen. Such units are portable and permit relatively rapid analyses.

In an Orsat analyzer, a sample of the gas mixture is trapped in a calibrated glass water-jacketed chamber, and the volume of the sample is accurately measured. The sample is then passed through an absorbing medium in one of three glass absorption pipettes where carbon dioxide is removed from the mixture. The gas sample is then returned to the measuring chamber where its volume is measured again. The difference between the volumes before and after the absorption gives the percentage of carbon dioxide absorbed by the medium. In similar fashion, the oxygen and carbon monoxide contents are determined in consecutive order in the other absorbing pipettes. Manipulation of the gases from one chamber to the other is accomplished by raising and lowering a water-leveling bottle and turning cocks in the gas connecting lines.

Another unit which operates on the Orsat absorption principle is available for measuring only carbon dioxide or oxygen, quickly and simply. It consists of a small calibrated plastic tube with a reservoir at top and bottom. The unit is "hour glass" shape. A gas sample is pumped in to fill the upper reservoir, while the absorbing liquid is in the bottom reservoir. The analyzer is then turned upside down, and the liquid flows down through the absorption tube. Then the unit is turned upright again for repeating the absorption cycle. Absorption of gas results in a partial vacuum, causing the column of liquid, supported on a diaphragm, to rise in the tube. Any change in height of the liquid column, before and after the manipulations, indicates by direct percentage reading the carbon dioxide or oxygen content of the gas sample.

36.12 Specific-Gravity Analyzers

The specific gravity of gases being tested can be measured and compared against that of air. Since carbon dioxide is much heavier than the other constituents in sintering gases, an instrument of this type is especially sensitive to changes in carbon dioxide. Therefore, the instrument is useful, for example, in checking sintering furnace atmospheres of purified rich exothermic or rich endothermic gases which are supposed to be carbon dioxide-free and dry. Should carbon dioxide appear in the influent or effluent furnace atmosphere gases, a pronounced specific gravity difference would show the need for corrective measures.

As another example, the inert gas used for purging air from a furnace at start-up, or flammable gas during shutdown and the flammable gases used for the atmosphere, are appreciably lighter (or heavier) than air. By continuously measuring the specific gravity of the effluent gas from the furnace, the completeness of purge can be determined. This is important, to prevent explosions and to avoid contaminating the furnace atmosphere with air or moisture in an incompletely purged furnace.

For automatically operating mixing or regulating valves, or other control mechanisms, a pneumatic transmitter can be installed in the instrument. The instrument can be equipped also with electronic contactors for signalling faulty conditions or, with interlocks, to prevent start-up or to create a shutdown if hazards exist.

The analyzer contains two motor-driven impellers located in separate chambers rotating in opposite directions. Facing each impeller is a companion impulse wheel. The shafts for these impulse wheels are connected by an external linkage. The action of each impeller and impulse wheel is somewhat similar to the fluid drive in automobiles except that the gas and air are substituted for the transmission oil.

A sample of gas to be tested is drawn into one chamber by an impeller; the rotation creates a torque on the impulse wheel which is proportional to the gas density. Atmospheric air is drawn into the other chamber, and in like manner results in a torque in the opposite direction on the other impulse wheel. The difference between these opposing torques in the impulse wheels is a measure of the specific gravity indicated by movement of a pointer on a scale.

36.13 Infrared Analyzers

Different gases absorb infrared energy at characteristic wave lengths. Because of this property, changes in the concentration of a single component in a mixture, produce corresponding changes in the total energy remaining in an infrared beam passed through the mixture. These energy changes, which are detected by an infrared analyzer are therefore a measure of the gas concentration. Proper selection of apparatus permits accurate, rapid analyses for such constituents as carbon monoxide, carbon dioxide, and methane. One make of instrument also measures water in terms of dew point. Each gas compound absorbs a certain portion of the infrared spectrum which no other gas absorbs, and the amount radiation absorbed is proportional to the concentration of the specific gas. Such instruments are not suitable for measuring oxygen, hydrogen, and nitrogen, which have no infrared absorption band.

Infrared analyzers have very high sensitivity and selectivity. An example is the measurement of 0 to 10 parts per million carbon dioxide in a 6-component hydrocarbon stream.

Typical applications for infrared analyzers are in the field of gases with high carbon potential, such as purified rich exothermic gas or dry endothermic gas, and those with high purity such as dry hydrogen or argon. Either water or carbon dioxide is a good indicator of carbon potential. Small changes in the amounts of these constituents, in some situations, can make wide changes in carbon potential. Also, differences in dew point of dry hydrogen or argon can affect sintering results on reactive metals. The high sensitivity of the infrared analyzer is of value, particularly for large furnaces with high output, where the substantial cost of this type analyzer can be justified. These analyzers also can be equipped to control the level of water or carbon dioxide to provide carbon-potential control.

36.14 Moisture Detectors

The moisture content of an atmosphere gas often has a direct bearing on results obtained in sintering. The most convenient means of determining moisture content is by checking the dew point. Sometimes it is necessary or desirable to know the dew point of a gas entering and leaving a refrigeration-type gas cooler or regenerative-type dryer. The dew point of a gas sample taken from a furnace chamber is frequently needed, in comparison with entering dew point, to determine the increase of water, if any. Excess water in a supposedly non-decarburizing atmosphere can cause decarburization. Expressed in another way, the water content usually is a good indicator of the carbon potential. Since carbon dioxide reacts with hydrogen at elevated temperatures to form water and carbon monoxide, by the water-gas reaction, an increase in the dew point of supposedly carbon dioxide-free atmosphere containing hydrogen will indicate the presence of carbon dioxide in that atmosphere as being the real source of the water. Also, water often causes oxidation of reactive metals, even when present in relatively small quantities in otherwise pure atmospheres. Therefore, the dew-point determination is a good all-round method of making a quick check to see whether conditions are right throughout a furnace system.

A relatively simple and inexpensive apparatus for indicating dew point is called a "dew cup." This device consists of an outer container with an observation window, through which the gas to be tested passes. Within the container is a cup, nickel-plated and highly polished on the outside. This cup is situated directly in the gas stream so that the entering gas will impinge on its polished surface. Within the cup is a glass thermometer. To make a test, the atmosphere gas is directed to flow through the container, and acetone is poured into the cup. After about 5 minutes, small amounts of crushed dry ice are added to the acetone, while stirring constantly with the thermometer. At the first sign of dew or moisture on the polished surface of the cup, the temperature is read from the thermometer. This reading is the dew point of the gas being tested.

Another type of dew point indicator, somewhat more complex but relatively simple to operate, is an instrument which involves compressing a sample of the gas, then quickly expanding it. If the gas, due to its rapid expansion, has cooled below its dew point, a fog will be observed in the expansion chamber. The procedure is repeated to find the end point or vanishing point of the fog or condensation. A reading is taken from a pressure-ratio gauge, which indicates directly the relation between the pressure of the gas sample at the end point and of the atmospheric pressure. By means of a separate dial calculator, this pressure ratio is converted to dew point.

The amount of water vapor in an atmosphere gas can be determined by its action on a hygroscopic salt. If dry lithium chloride salt, for example, is exposed to the atmosphere gas, it will absorb moisture and dissolve, forming a saturated solution. If the salt and the saturated solution are heated, the water in the solution will evaporate. At equilibrium temperature the vapor pressure of the water in the solution is equal to the partial pressure of the water vapor in the atmosphere gas. This temperature is a measure of the partial pressure of water vapor in the atmosphere gas—that is, the dew point temperature.

The humidity-sensitive element is a thermometer bulb inside a thin-walled metal tube covered with a woven-glass tape impregnated with lithium chloride. A pair of gold-overlay wires is wound over the tape, and surrounded by a perforated metal guard. When the salt absorbs moisture from the gas, it becomes an electrical conductor. Current is passed between the two wires to raise the temperature of the element until equilibrium is attained. This temperature is sensed by the thermometer bulb inside the tube and transmitted to a recorder calibrated to read the dew point of the gas directly.

Application of this method is useful for indicating, recording, and even controlling dew points in sintering furnaces in which carbon potential (dew point) is critical. The system is particularly useful for gases entering the furnace. Control is achieved by automatically adjusting the ratio of the air and gas entering the gas producer. If contaminants, such as distillation products from lubricants in the work, are negligible in the effluent gas from the furnace, the system can be extended further to control the effluent dew point. This is quite desirable, since effluent gas is much more representative than influent gas of actual conditions surrounding the work. In this case, fastest response generally is achieved by injecting and regulating a small amount of hydrocarbon gas, such as methane or propane, directly into the heating chamber. The dew point controller automatically regulates a valve in the auxiliary gas line to hold a constant preset dew point within the furnace, and thereby maintains uniform carbon potential.

Another useful type of dew point indicator and recorder is the infrared type of one particular make, as discussed under Infrared Analyzers (36.13).

36.15 Carbon Potential Control

The carbon potential of a furnace atmosphere may be defined as its carbon equilibrium with steel at a specific temperature. For example, to sinter an iron-graphite mixture and maintain a 0.80% carbon in the resulting steel at 2050F (1121C), the furnace must have a carbon potential of 0.80% C at the sintering temperature of 2050F (1121C). Either dew point (water vapor) or carbon dioxide content can be used as a measure of carbon potential of an endothermic-type sintering atmosphere. To put into successful practice a carbon-potential control system, one must understand: (a) Sampling, (b) Measurement, (c) Control, and (d) Effect of various temperature zones on Control.

a) Sampling: The gas sample taken for measurement of the carbon potential should be taken from the high-heat, sintering soak zone of the furnace. A sample tube of a heat resisting alloy or a refractory tube of 1/4 in. (6 mm) ID will serve this purpose.

For sintering temperatures up to 2100F (1150C), a heat-resisting alloy tube, such as Inconel, is more convenient to install gas-tight than a refractory tube. If the furnace is of a non-muffle type, the tube must extend at least two inches into the hot zone beyond the refractory wall. If the furnace is of the muffle type, the sample tube must be welded to the muffle and clearance allowed in the refractory brick and shell of the furnace for expansion of the muffle. It is very important to connect a tee at the end of the sample tube that extends outside the furnace shell, so that a valve or clean-out plug can be connected for daily cleaning with a wire brush or by blowing compressed nitrogen through the tube to remove carbon deposits. Carbon deposits in the sampling tube will cause erroneous dew point or carbon dioxide readings. Copper or aluminum tubing at 3/8 in (9.5 mm) OD can be connected to the 90° outlet of the tee on the gas sample tube. A filter should be installed about 15 in (380 mm) from the outlet of the gas sample tube to prevent the extension tube from becoming contaminated between the furnace and the instrument. Another filter must be installed immediately ahead of the instrument to assure that no dirt or condensation enters the measuring instrument.

Since the furnace is not under positive pressure, a pump is required to pump a gas sample from the furnace through the instrument. Because most instruments are flow-sensitive, it is essential to follow the instrument manufacturer's recommendations on the flow requirements. A small flowmeter usually is used to measure the required gas sample, and to maintain consistent results from the instrument. The endothermic generator operates under pressure, and thus a sample pump is not required for it. The same recommendations of double filters, however, and a sample measurement flowmeter apply.

b) Measurement: Instruments to measure a gas constituent to control the carbon potential have been described above. These instruments are for either dew point (moisture control) or carbon dioxide measured by infrared. The instruments can be indicating with a manual control, or indicating or recording with full automatic control. The choice of the dew point type or the infrared carbon dioxide type is a matter of economics and application. The dew point instruments are less costly than the carbon dioxide infrared analyzers. For single-point control, both will do a satisfactory job. Dew point analyzers are not as sensitive, and are slower in response than carbon dioxide infrared analyzers. This may or may not be a disadvantage. Because of the fast response, infrared analyzers are used for multi-point control systems. For example, when one analyzer is used to measure and control the generator, the sintering furnace, and the hardening furnance, the infrared carbon-dioxide analyzer is used because of the speed of response and sensitivity of measurement. Automatic valves are used to divert the sample gas from the generator to the sintering furnace and to the hardening furnace at one-minute intervals. Thirty seconds are allowed to purge the instrument with the new process gas sample, and thirty seconds are then allowed for the instrument to analyze the gas and to send the signal to the controller to make the proper adjustment to bring the process under control. When infrared analyzers are used for multi-point control the cost can then compete, per point of control with dew-point analyzers. Dew-point is not satisfactory to use for multi-point control because water molecules are difficult to purge from an instrument in a short time.

c) Control: Whether the process is to be controlled manually or automatically, the same principle applies. For carbon potential control of an endothermic-type atmosphere, both the generator and the furnace should be controlled to achieve the best results. If the process is for either sintering or hardening of P/M parts where it is desired to control the carbon on the part above the 0.50% carbon range, the generator should be controlled at 25 to 30F (-3.9 to -1.1C) dew point or 0.25 to 0.30% carbon dioxide.

To control the process, whether it be sintering or hardening, it is necessary to refer to equilibrium data of dew point vs. carbon content, or carbon dioxide vs. carbon content, at the operating temperature of the process. Suppose one desired to sinter an iron-1% graphite mixture at 2050F (1121C) and maintain a 0.80% carbon content in the resulting sintered steel. Referring to figure 36-1, the equilibrium relationships between dew point and percent carbon in plain carbon steels at various furnace temperatures, shows that a 15F (-9.4C) dew point is required. Since the endothermic gas entering the furnace will be between 25 to 30F (-3.9 to -1.1C), the dew point will be much higher than shown by the equilibrium data. To achieve the 15F (-9.4C) dew point in the furnace, natural gas or another hydrocarbon gas such as propane, is added to the endothermic gas between the generator and the furnace. The mixture is put into the hot sintering zone only, and at several inlets. The hydrocarbon gas reacts with the water vapor and carbon dioxide constituents in the endothermic gas in the furnace at 2050F (1121C) to reduce the decarburizing constituents to the desirable level, as shown by the equilibrium data. If the process is controlled manually, the hydrocarbon gas is added in small increments through a needle valve and measured by a small precision flowmeter. If the process is automatic, the control instrument sends a signal to a motorized or pneumatic valve to add the proper amount of hydrocarbon gas to achieve the desired control point. It is very important that the gas sample tube be not adjacent to the gas inlet tube. The endothermic gas and the hydrocarbon gas must be fully reacted before being withdrawn from the furnace to be analyzed by the instrument.

If the process to be controlled is to sinter low-carbon iron parts requiring dew points higher than the 25 to 30F (-3.9 to -1.1C) generator dew point, this can be achieved either by controlling the generator output at a higher dew point, or by adding air to the endothermic atmosphere instead of a hydrocarbon gas. In some instances where both high and low dew points are to be controlled in the furnace, automatic controllers are designed to add either air or hydrocarbon gas, as the process dictates.

The equilibrium data in Figure 36-1 were determined by actual measurement, and are fairly accurate. However, the data should be used only as a guide to tune the process. The final adjustment should be made after the carbon analysis of the P/M sintered part is determined. When controlling carbon dioxide, it is recommended that a small portable indicating dew point meter be used in conjunction with Figure 36-1 to tune in the process. Then the corresponding carbon dioxide reading on the infrared unit should be used as the control point. One can then make his own carbon dioxide vs. carbon content equilibrium diagram for his own particular furnace.

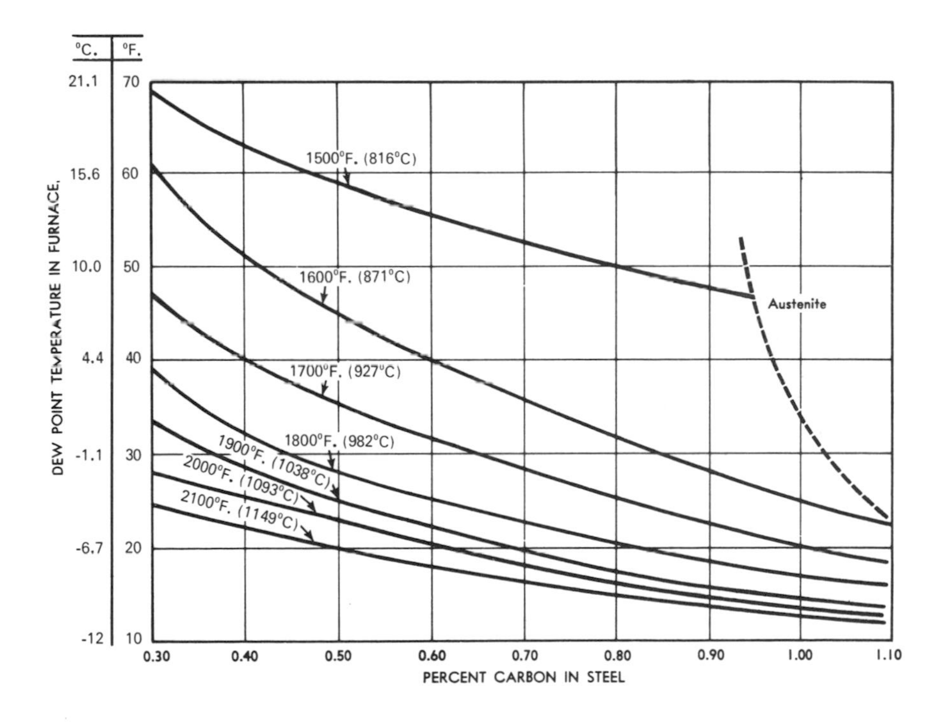

FIGURE 36-1

Dew point and carbon equilibrium diagram.

d) Effect of various temperature zones on Control: Referring to Figure 31-2, there are various temperature zones in a continuous sintering furnace. To achieve a specific carbon content in the sintered product, it is important to understand the effect of temperature of each zone on the final carbon equilibrium of the P/M part. The burn-off or preheat section, and possibly the first zone of the high-heat section, will run at a lower temperature than the final high-heat sintering zone. By referring to the equilibrium diagram in Figure 36-1, one will note that if the carbon potential of the atmosphere is adjusted for the 2050F (1121C) high-heat zone, an excessive amount of carbon will be picked up at the lower temperatures. This is no problem in reference to the final desired carbon in the P/M part, as the excessive carbon will be extracted in the high-heat zone to the proper equilibrium point. However, any lower-temperature zones following the high-heat zone will add carbon, as shown by the equilibrium diagram. Thus excessive carbon over the desired amount will remain after cooling.

This is of no great concern when one wishes to maintain the maximum carbon content of 0.80% to 1.00%. Geneally speaking, it is difficult to maintain these levels even with the best of sintering furnaces and atmosphere controls, and the build-up of carbon in the insulated cooling zone will not exceed the desired high carbon content for the short time the parts are exposed to the high carbon potential atmosphere at the lower temperature.

It may be a problem, however, to control the carbon at some exact equilibrium point under the maximum of 0.80 to 1.00%, say 0.30 or 0.50%, to overcome the effect of the varying lower-temprature range of an insulated pre-cooling zone. Several things may be done to achieve a semblance of control. First, eliminate the insulated cooling zone so the parts will cool rapidly as they enter the water-jacketed cooling zone. This may reduce the time sufficiently in the transition-temperature range to avoid carbon pick-up. Eliminating the insulated cooling zone, however, may cause higher maintenance on the first water-jacketed chamber, and may also cause too much thermal shock on the P/M parts being sintered.

The second solution is to dilute the endothermic gas in the transition zone with a gas that has no carbon potential, such as dissociated ammonia or nitrogen. For example, two-thirds of the total atmosphere-gas flow can be enriched endothermic gas to achieve the desired carbon potential put into the high-temperature sintering zone, and one-third dissociated ammonia can be put into the intermediate-temperature insulated pre-cooling zone. By trial and error, the exact proportions can be worked out to achieve the desired surface carbon chemistry.

Another solution is to put a sample tube into the insulated pre-cooling zone and add another control point to the instrument. A lean, unpurified, exothermic gas can be added to the unenriched endothermic gas supplied to this zone, if necessary, by the controller to achieve a dew point or carbon dioxide level to be in approximate equilibrium with the P/M parts in the varying-temperature transition zone. The water vapor or carbon dioxide set point for control will have to be worked out by trial and error. Once the point is determined for a specific production rate, automatic control can be achieved continuously.

36.16 High-Vacuum Gauges

It is very important in vacuum powder metallurgy sintering to know the pressure in heating chambers and perhaps in headers, manifolds, and at the pumps. A variety of vacuum gauges is available. Some of the more common ones will be mentioned here to acquaint you with their names, operating principles, and pressure ranges.

The McLeod gauge is a mercury manometer of such precision that is is often used for checking the calibration of other types of vacuum measuring and control instruments. By manipulating the McLeod gauge and measuring the difference in height of two mercury columns in adjacent capillary tubes, it is possible to read the pressure of the system in microns of mercury on a calibrated scale. In the manipulation, the measured sample is compressed to approximately atmospheric pressure by the column of mercury in one dead end capillary. Any vapors, such as moisture, will condense under this pressure and introduce an error. The gauge gives true readings only with dry permanent gases such as nitrogen, oxygen, and air. Some portable McLeod gauges incorporate an expendable dessicant tube to trap moisture before it enters the gauge. Indicating crystals show when the dessicant should be discarded or dried. McLeod gauges are built to cover various ranges. One covers from about 5.0 mm to 0.01 microns Hg (665 to 1.33 Pa); another about 2 microns to 1×10^{-7} mm Hg (0.266 to 0.0133 Pa).

The Pirani gauge employs a Wheatstone bridge circuit to balance the resistance of a resistor or of a tungsten filament, sealed off in high vacuum against that of a tungsten filament which can lose heat by conduction to the gas being measured. A change of pressure causes a change in the filament temperature, and, consequently, of the wire resistance, thus unbalancing the bridge. The amount of this unbalance is indicated on a microammeter and read in terms of pressure. Typical Pirani gauges cover a range of about 2.0 mm to 1 micron Hg (266 to 0.133 Pa) but one type covers the unusually broad operating range of about 100 mm to 1 micron (133×10^3 to 133×10^{-3} Pa).

The thermocouple gauge consists of a thermocouple connected to a tiny heating unit in a sensing tube placed in the vacuum system. A constant current through the heater produces a given temperature, resulting in an EMF (electromotive force) across the terminals of the thermocouple which is fed through a calibrating potentiometer to a simple meter for pressure indication. As the pressure is reduced, so is the cooling effect of the surrounding gas, resulting in an increase in the temperature of the thermocouple. Such variations in temperature are thus used to indicate variations in pressure in the system. Thermocouple vacuum gauges have ranges of about 1.0 Torr to 0.001 Torr (133 to .133 Pa).

Ionization gauges are of two types: cold-cathode (ordinarily called "Philips" or "discharge" gauges), and hot-filament (usually simply called "ionization" gauges). The principle of operation is that collisions between molecules of gases and electrons result in formation of ions. Below about 1 micron Hg this formation of ions varies linearly with pressure. Measurement of this ion current can be translated into units of gas pressure.

In the cold-cathode gauge, the ion current is produced by a high-voltage discharge. The cathode and anode are suspended in a magnetic field in the sensing element. The electrons emanating from the cathode are caused to spiral as they move across the magnetic field to the anode and they ionize gas molecules in their paths. The ionization current is directly proportional to the molecular density, and when passed through a microammeter it gives a pressure indication. The Philips gauge has a wide operating range, a typical one being from 25 microns 0.01 microns Hg (3.325 to 1.33×10^{-3} Pa).

In the hot-filament ionization gauge, the electrons emitted from a heated filament in the sensing tube are accelerated toward a positively-charged cylindrical grid. Some electrons pass into the space between the grid and a negatively-charged collector and collide with gas molecules from the vacuum system to produce positive ions. Here again the ion current is proportional to the number of molecules and to the pressure which is read on a microammeter. Hot-filament gauges have several ranges, all starting with a maximum of 1 micron Hg (0.133 Pa), with typical minimums of about 10^{-2} to 10^{-9} microns (0.133×10^{-2} to 0.133×10^{-9} Pa).

Another type of ionization gauge is the Alphatron gauge which utilizes a radioactive source to produce alpha particles that ionize the gas molecules. This radioactive source replaces the filament, or electron source, in the hot-filament gauge. The range of the Alphatron gauge is wide—in one typical case from 1000 Torr to 0.0001 Torr (0.1 micron Hg or 0.0133 Pa).

To summarize, the McLeod, Pirani, and thermocouple gauges are used down to pressures of about 1 micron Hg, the Alphatron gauge to 0.1 micron Hg, and the ionization gauges starting with a top pressure of 1 micron and going down to extremely low pressures, in one case to 2×10^{-12} Torr. The actual pressure ranges vary somewhat with type and make, and the ranges given here are just typical.

37.00 EFFECTS OF FURNACE ATMOSPHERES ON HEATING ELEMENTS

It is very difficult to compare the life of heating elements in various furnace atmospheres in terms of expected hours of operation. The life of a resistance material depends not only on furnace atmosphere, but also on many other working conditions, i.e., watt density loading, cross-sectional area, temperature of operation, frequency of switching the current on and off, support of the resistor in the furnace, physical shape and design of the resistor, and how hard the furnace is worked.

The experienced furnace designer knows the shortcomings of a particular resistor to a specific furnace atmosphere, and designs more conservatively than he would for an atmosphere that has little effect on the resistor material. Thus, a comparison in hours for the same resistor in commercial furnaces operated at the same temperature would not give a true picture of the action of the different atmospheres. The conservative design would not be used for the more favorable atmospheres because of economic considerations. Therefore any data available on hours of operation would be misleading and very controversial in nature.

To help select the proper resistor material as a heating element in a known furnace atmosphere Table 37-I was prepared. The table is based on the recommended furnace operating temperatures for typical heating-element resistor materials for the specific atmospheres. This gives the engineer a choice of materials, depending upon the expected maximum operating temperature of each furnace. Resistor temperatures are always higher than furnace control temperatures, and the difference depends upon the design of watt density loading of the resistor. Therefore, when the furnace is to be operated close to the maximum temperature, the watt density loading must be lower and more conservative. If this rule is kept in mind, plus the rule that the cross-sectional area of the heating element should be increased for furnace atmospheres that show more attack than others, Table 37-I will serve as guide to the proper selection of the resistor material for a specific application, and should lead to a workable design. To supplement Table 37-I, the following brief discussions of the action of various furnace atmospheres on heating-element resistor materials may be helpful.

37.11 Oxidizing Atmosphere

With the exception of air sintering for aluminum P/M parts little if any sintering is done in air atmospheres. However, the atmosphere outside metal muffles or retorts is often air, so some comments regarding heating elements in air atmospheres may be helpful.

With the exception of molybdenum, tungsten and graphite, all other

TABLE 37-I
MAXIMUM RECOMMENDED FURNACE OPERATING TEMPERATURES(a)

Furnace Atmosphere Gases(b)	Air: Natural Gas Ratio	Dew Point °F Entering Furnace	Heating Element Materials (g). Temperature Indicated in °F					
			35Ni-20Cr-45 Fe	80Ni-20Cr	Moly	Tungsten	Silicon Carbide	Graphite(c)
1 Hydrogen								
A. By Electrolysis of Water								
1. Direct from Cells								
a. Unpurified—Saturated		+70 to +90	1800	2150 (d)	3200	3200	2350	NR
b. Purified		-80 or lower	1800	2050 (i)	3400 (i)	4500 (i)	CR-2100 (i)	4000
B. Bulk Compressed Gas								
1. Unpurified		-30 to +20	1800	2050 (d)	3200	3200	2350	NR
2. Purified		-80 or lower	1800	2050 (i)	3400 (i)	4500 (i)	CR-2100 (i)	4000
C. By Catalytic Conversion of Hydrocarbons								
1. Unpurified—Saturated		+70 to +90	1800	2150 (d)	3200	3200	2350	NR
2. Dried		-100 or lower	1800	2050 (i)	3400 (i)	4500 (i)	CR-2100 (i)	4000
D. From Liquid Hydrogen		-84 or lower	1800	2050 (i)	3400 (i)	4500 (i)	CR-2100 (i)	4000
2 Nitrogen—Bottled		-64 or lower	1800	2100	3000 (k)	3000 (k)	2400	5000
3 Hydrogen-Nitrogen Mixtures								
A. Dissociated Ammonia								
1. As Reacted—Dry		-60 to -40	1800	2050 (i)	3200 (i)	4500 (i)	CR-2100 (i)	4000
2. Moisture Added—Saturated		+70 to +100	1800	2150 (d)	3000 (k)	3000 (k)	2350	NR
B. Burned Dissociated Ammonia								
1. Rich—Saturated		+70 to +90	1800	2150 (d)	3000	3000	2350	NR
2. Lean—Saturated		+70 to +90	1800	2100	3000 (k)	3000 (k)	2350	NR
C. Direct Catalytic Conversion of Ammonia and Air								
1. Rich								
a. As Reacted—Saturated		+70 to +90	1800	2150 (d)	3000	3000	2350	NR
b. Cooled to 40°F		+40	1800	2100 (d)	3000	3000	2350	NR
c. Dried		-80 or lower	1800	2050 (i)	3200 (i)	4500 (i)	CR-2100 (i)	4000
2. Lean								
a. As Reacted—Saturated		+70 to +90	1800	2100	3000 (k)	3000 (k)	2350	NR
b. Cooled to 40°F		+40	1800	2100	3000 (k)	3000 (k)	2350	NR
c. Dried		-80 or lower	1800	2100 (i)	3200 (i).(k)	4100 (i).(k)	CR-2100 (i)	4000
4 Reformed Hydrocarbon Gases								
A. Exothermic Gas								
1. Rich								
a. As Reacted—Saturated	6:1	+70 to +90	1800	2150 (d)	NR	NR	2350	NR
b. Refrigerated	6:1	+40 to +60	1800	2100 (d)	2500	2500	2350	NR
2. Lean—Saturated	10.25:1	+70 to +90	1800	2100 (d)	NR	NR	2350	NR
B. Purified Exothermic Gas								
1. Rich	6:1	-40 or lower	1800	2000	NR	NR	2350	4000
2. Medium Rich								
a. As Reacted	6.75:1	-40 or lower	1800	2000	NR	NR	2350	4000
b. Methane Added	6.75:1	-40 or lower	NR	CR (e)	NR	NR	2350 (h)	4000
3. Lean	10.25:1	-40 or lower	1800	2100	NR	NR	2350	4000
C. Endothermic Gas								
1. Rich—Dry	2.4:1	-10 to +10	CR-1800	2000	NR	NR	2350	NR
2. Rich—Fairly Dry								
a. As Reacted	2.6:1	+20 to +30	1800	2000	NR	NR	2350	NR
b. Methane Added	2.6:1	+20 to +30	NR	CR (e)	NR	NR	2350 (h)	NR
3. Medium Rich—Saturated	3.5:1	+70 to +90	1800	2150 (d)	NR	NR	2350	NR
4. Lean—Saturated	4.5:1	+70 to +90	1800	2150 (d)	NR	NR	2350	NR
5 Argon—Bottled		-73 or lower	1800	2100	3200 (k)	4100 (k)	2600	5000
6 Helium—Bottled		-73 or lower	1800	2100	3200 (k)	4100 (k)	2600	5000
1-6 Above Atmospheres Containing Impurities of:								
A. Sulphur								
1. Up to 10 Grains/C cu. ft.			CR-1800	CR-2050	NR	NR	Temps. Same as Above (f)	NR
2. 10 to 25 Grains/C cu. ft.			Cr-1800	NR	NR	NR	Temps. Same as Above (f)	NR
B. Lead			NR	NR	NR	NR		NR
C. Zinc								
1. From Brass, Openly Loaded			CR	CR	NR	NR	Temps. Same	NR
2. From Zinc Stearate			1800 (h)	2050 (h)	NR	NR	as Above (h)	NR
7 Vacuum (Below 150 Microns)			NR	1850 (j)	3300 (j)	4500 (j)	2100 (j)	4000 (j)
8 Air (g)								
A. Normal		+50 or lower	1700	2100	NR	NR	2600	NR
B. Wet		Above +50	1700	2100	NR	NR	CR	NR

Footnotes to Table 37-I

(a) This information serves only as a guide. It is not to conflict with nameplate data, operating instructions, or bulletins supplied with or related to any furnace. Data pertain only to heating elements. Lower maximum temperatures often are necessary due to limitations of refractories and supports.

Temperatures listed for all materials, except graphite can be taken as furnace temperatures, provided elements are designed with conservatively low watts release for good life at maximum operating temperature.

Data given are based on typical heating elements in general use for powder matallurgy as of November 1, 1976.

(b) For approximate compositions and costs of atmosphere gases see Table 19.1.

(c) Element temperatures (not ambient) commonly used in controlling graphite tube furnaces. In most cases life is measured in weeks, but is considered acceptable in order to achieve the high temperatures indicated.

(d) Due to "green rot" failure by wet H_2 and/or CO (but not when dry) life may be reduced substantially at temperatures below 1800° F. The 35 Ni-20 Cr Alloy is not subject to "green rot" and is preferred to 80 Ni-20 Cr below 1800° F.

(e) Special 80 Ni-20 Cr elements with ceramic protective coatings designed for low voltage (8-16 volts) operation can be used successfully.

(f) Use 2400° F maximum for argon and helium with sulphur impurities instead of 2600° F maximum listed above.

(g) Elements not listed above, useful primarily in air, are as follows:

(1) 23Cr-4.5AL-2 Co-Bal. Fe—Good to 2100° F in air. Fair to 2100° F, if oxidized prior to admitting H_2 atmosphere.
(2) 24Cr-6.0Al-2 Co-Bal. Fe—Good to 2400° F, if oxidized prior to admitting H_2 atmosphere.
(3) 37Cr-7.5Al-Bal. Fe —Good to 2400° F, if oxidized prior to admitting H_2 atmosphere.
(4) Platinum —Good to 2550° F in air.

(h) When heating elements and walls are exposed to deposits of carbon and zinc from volatilized lubricants, or to carbon from ethane and propane, these deposits combine with, deteriorate, and short circuit heating elements, and overload control and power supplies. Such deposits should be burned out with air frequently at 1200 to 1400° F, to help element life and preserve the equipment. Carbon combined with heating elements or brickwork cannot be "burned out," so diligence is required to prevent or minimize its presence on and in them. This becomes increasingly important with increasing dryness at atmospheres. Moisture in the atmosphere reacts with the volatiles and helps to minimize or "clean up" such deposits. Dry atmospheres doe not give this protection.

(i) Data based on assumption that precautions will be taken to assure that high-purity atmosphere will be maintained within furnace chamber surrounding heating elements. For example, for dry-hydrogen applications, it is assumed that the dew point of effluent gas from the heating chamber will be about −40° F or drier. The sample must be taken at a distance from the inlet so as to include contaminents.

(j) Vacuum furnace maximum temperatures are only conditionally recommended, and life may be short or long dur to variable design and operating conditions including. Time-temperature-pressure cycles; reactions with various contaminating gases during degassing periods at different temperatures; compositions, volatilization and creep rates of element materials; progressive mechanical distortion of heating elements within the plastic deformation range; recrystallization under heat; and mechanical abuse during loading and unloading.

(k) Precautions are necessary to prevent oxygen contamination of atmosphere by infiltrating air, since free oxygen can cause damaging oxidation of moly or tungsten heating elements. Such protection can be had by using purging chambers, or by mixing a small amount of hydrogen with the atmosphere gas to react with oxygern to form H_2O and thereby protect elements from oxidizing. In the latter event, dryness of the furnace atmosphere to satisfy footnote (i) would be destroyed, and maximum temperature should be about 3000° F.

NR-Not recommended as one of the best element materials for this atmosphere.

CR-Conditionally recommended where shorter life is acceptable. For recommendations consult furnace manufacturer.

Conversion Chart

F	C
−100	−73
−90	−68
−80	−62
−70	−57
−60	−51
−50	−46
−40	−39
−10	−18
+10	−12
20	−7
30	−1
40	+4
50	10
60	16
70	21
80	27
90	32
100	38
—	—
1750	955
1850	1010
1950	1065
2050	1120
2150	1175
2250	1230
2350	1290
2600	1426
3000	1650
3200	1760
3400	1870
4000	2200
4500	2480
5000	2760

resistor materials commonly used have maximum life when operating in an oxidizing atmosphere of air. Air forms an impervious, oxide protective film to reduce further attack by oxidation. The protective oxide coating produced by air on the nickel-chromium series is so impervious that the base metal in the core of overheated elements may melt and run out at the overheated sections, leaving the oxide skin of the original shape of the heating element intact on the walls of the furnace. The inexperienced operator often claims the failure is due to a hollow defect in the resistor material, instead of a defect in the temperature control causing overheating.

Since the 80 nickel-20 chromium resistors have the greatest resistance to oxidation, and can be used to operate furnace temperatures up to 2100F (1150C), this is the alloy most commonly used in air atmospheres outside muffles and retorts.

The chromium-aluminum-iron alloy series, which may or may not contain a small percentage of cobalt, are used primarily in the temperature range of 2100 to 2400F (1150 to 1315C) in air.

For element temperatures up to 2900F (1590C) in air, the non-metallic, silicon-carbide type is most common. The maximum life of the resistor is obtained when operating in an atmosphere of dry rather than moist air. This resistor is more commonly used than the chromium-aluminum-iron type. Molybdenum-disilicide elements, in relatively small sintering furnaces, are used up to 2900F (1590C) in air. Because of the high cost of platinum, it is used only in very special cases, where silicon-carbide cannot be worked into the furnace design. Platinum is restricted to air, and cannot be used in reducing atmospheres.

37.12 Reducing Atmosphere

The term "reducing atmosphere" refers to the atmospheric reaction with iron and iron oxide.

Reducing atmospheres may be divided into four categories: (1) dry hydrogen or dissociated ammonia containing no oxidizing or carburizing constituents; (2) unpurified rich exothermic atmosphere having a low carbon potential or decarburizing in nature; (3) purified rich exothermic atmosphere having medium carbon potential; and dry endothermic or charcoal atmospheres having a high carbon potential; (4) dry endothermic atmosphere enriched with a hydrocarbon gas to produce a carburizing atmosphere, or enriched with ammonia and hydrocarbon to produce a carbonitriding atmosphere.

With the exception of dry hydrogen or dissociated ammonia, all the above atmospheres are oxidizing to the nickel-chromium series. Even the hydrogen or dissociated ammonia atmospheres will oxidize chromium, unless the gas is extremely dry. The type of oxide produced by the so-called reducing atmosphere is entirely different than that produced by air. The oxide produced by air is a green to black impervious type that retards further oxidation of the metal below its surface. The oxide produced on high-nickel-chromium elements by wet reducing furnace atmospheres is green and not impervious, and the atmosphere keeps attacking the base metal. This type of attack has generally been referred to as "green rot." It occurs in some alloys such as 80Ni-20Cr in wet hydrogen, rich exothermic and wet endothermic gases in a limited temperature range of 1650 to 1850F (900 to 1010C) element temperature. The 35Ni-20Cr series with at least 1.25% silicon, however, is quite resistant to "green rot" attack. Accordingly, 35Ni-20Cr is recommended in wet hydrogen-bearing gases up to 1800F (980C) furnace temperature, and 80Ni-20Cr above 1800F (980C).

However, using a columbium stabilized 80Ni-20Cr alloy (1.25 Cb) will, for most practical applications and atmosphere conditions, eliminate or greatly minimize effects of "green rot."

Dry hydrogen and dissociated ammonia have the least affect of the listed reducing atmospheres on the nickel-chromium group. At temperatures above 2000F (1090C), the resistor may have better life in dry hydrogen than in air, because the oxidation rate in air becomes more rapid at elevated temperatures.

Above 2000F (1090C) furnace temperature the unpurified rich exothermic atmosphere has less harmful affect than the higher carbon potential purified rich exothermic, dry endothermic, or charcoal atmospheres on 80Ni-20Cr. The higher carbon potential atmospheres carburize the nickel-chromium alloys, especially at the higher temperatures. Chromium is a strong carbide-former, and may pick up enough carbon to lower the melting point of the alloy and cause localized melting and fusion in the heating temperature of 80Ni-20Cr to 2000F (1090C) maximum, when operating in reducing atmospheres of high carbon potential, unless the voltage can be reduced to lower the element temperature. Below 1900F (1040C) the 35Ni-20Cr alloy is best, despite the carburizing affect which does not necessarily mean that the effectiveness of the heating element is impaired.

Carburizing atmospheres produced by adding a hydrocarbon to purified rich exothermic gas, and dry endothermic gas, or by adding ammonia as well as a hydrocarbon to produce a carbonitriding atmosphere are not recommended for use on unprotected heating elements made of the nickel-chromium alloys. The short life of the resistor in this type of atmosphere is due to the carburizing effect, and also to shorting and melting of the leads and terminals because of carbon depositing on and becoming impregnated in the refractory lining. A special nickel-chromium heating element has been developed for operating in a carburizing atmosphere by protecting the alloy with a ceramic, high temperature enamel to resist carburization. The heating element operates on low voltage (8-10 volts), to prevent arcing at the terminals in the carbon-impregnated brickwork.

The chromium-aluminum-iron resistor alloys cannot be used successfully in any reducing atmosphere. Some success may be had in dry hydrogen if the alloy is previously oxidized in air, but the results are questionable. The chromium-aluminum-iron series depends upon the impervious oxide coating produced by air for its protection by high temperature oxidation, and therefore reducing or carburizing atmospheres are not recommended.

Molybdenum has excellent life up to 3400F (1870C) in dry hydrogen, dissociated ammonia, low-hydrogen mixtures of nitrogen, argon, or helium, or in rich exothermic gas of 60F (16C) dew point or less with 2500F (1370C) maximum operating temperature. In those gases, where hydrogen is absent or in minute quantity, a little hydrogen addition is considered desirable to help avoid oxidation of molybdenum by infiltrating air. Poor life results because of carburization when molybdenum is operated in a high carbon potential atmosphere. Air or high moisture will oxidize the elements at elevated temperatures. Molybdic oxide (MoO_3) is volatile when formed at high temperatures, and thus no protective oxide coating adheres to the molybdenum. If the proper atmosphere is not applied, the elements will volatilize and fail. When started up, the furnace should be purged with non-flammable atmosphere and then with hydrogen before turning on the power. In shutting down, protective atmosphere should be maintained down to about 700F (370C) to avoid harmful oxidation of the molybdenum by air.

Graphite resistors have been used for high-temperature furnace applications in atmospheres free of oxidizing constituents of oxygen, carbon dioxide and water.

Silicon-carbide resistors have good life in some reducing atmospheres. However, the maximum operating temperature is reduced from that when operating in air. Silicon-carbide elements also have the advantage that they can be operated in high carbon potential atmospheres that would easily carburize and destroy nickel-chromium elements at high sintering temperatures. The carbon deposited on the silicon-carbide resistors by stearate carry-over will eventually cause the resistors to draw more current, but this can be seen on the ammeter of the transformer, and the voltage to the resistors can be dropped easily by setting the transformer to a lower voltage tap until time permits the carbon to be burned out of the furnace.

37.13 Atmospheres Containing Contaminants of Sulphur, Lead or Zinc

Sulphur, if present, will appear as hydrogen sulphide in reducing atmospheres and as sulphur dioxide in oxidizing atmospheres. The source of sulphur contamination is usually due to high sulphur content of fuel gas used for producing the protective atmosphere, which generally can be eliminated, high-sulphur-bearing refractories or the material being processed in the furnace.

Sulphur is very poisonous to the nickel-chromium group of resistors in reducing atmospheres, and causes pitting and blistering of the resistor alloy in oxidizing atmospheres. The higher the nickel content of the resistor, the greater the attack will be. Therefore, if sulphur contamination is present and cannot be eliminated, the 35Ni-20Cr alloy will give the best performance.

Lead and zinc contamination result from the work being processed in the furnace. This is common in sintering furnaces for processing powder metallurgy parts. In the presence of a reducing atmosphere, lead vapor will leave a lead-bronze mix and contact the heating elements. Lead metallic vapors are even more poisonous to the nickel-chromium group than sulphur, and can damage the resistor in a matter of hours, under certain conditions of concentration and temperature. The higher-nickel alloys are more affected than the lower nickel alloys.

Zinc contamination results from zinc stearate used as a lubricant when pressing P/M compacts. The zinc stearate volatilizes out of the compacts when heated. Zinc vapors alloy with nickel-chromium heating elements, resulting in poor life. To eliminate the zinc vapors at the higher sintering temperatures, the common practice is to use a burn-off chamber operated at about 1200F (650C), as previously discussed. If this cannot be done, or if zinc is present in the metal being processed, then non-metallic silicon-carbide resistors should be used. Silicon-carbide resistors are not affected by sulphur, lead or zinc contamination, and therefore are commonly used at low or high temperatures, when contamination is anticipated.

37.14 Vacuum Atmosphere

For applications involving vacuum, 80Ni-20Cr has been used successfully up to 1850F (1010C) when the element is directly exposed to a low pressure. In furnaces where the elements are external to the low pressure chamber, or where the elements are exposed to a pressure above 100 microns (13.3 Pa), 80Ni-20Cr has been used successfully up to 2100F (1150C).

The 80Ni-20Cr alloy is not successful above 1850F (1010C) in low pressures, because the vapor pressure of chromium is high enough at elevated temperatures to remove the chromium from the resistors, resulting in poor life, contamination of the furnace and material being processed, and loss of vacuum. Because of this, the watt density loading at low pressures must be kept low and very conservative, especially at the higher operating temperatures. For furnace temperatures from 1850 to 3300F (1010 to 1815C), molybdenum is the best choice and gives satisfactory life. Tungsten and tantalum have been used successfully for furnace temperatures up to 4500F (2480), silicon carbide up to 2100F (1150C) at 3 microns pressure. Solid graphite, rod or plate elements are being used up to about 4000F (2200C) element temperature.

Graphite cloth elements for both heat treating and sintering of P/M parts are now available and are being used over a wide temperature range; 1600 to 3100F (870 to 1705C).

When a vacuum sintering furnace which is constructed of fibrous graphite felt and cloth is heated, a chemical reaction occurs between the graphite and the residual gases. This reaction alters the relationship between the partial pressure of the oxidizers and the total pressure of the system. Although the total pressure may remain stable, that is in quantity, the quality of the atmosphere improves due to the thermo-chemical reaction.

Specifically, the threshold of oxidation of graphite in air, water vapor, or carbon dioxide is

$$O_2 = 750\text{F} \ (399\text{C})$$
$$H_2O = 1350\text{F} \ (732\text{C})$$
$$CO_2 = 1850\text{F} \ (1010\text{C})$$

Therefore, when a furnace containing a residual atmosphere of these combined gases is heated above these oxidation thresholds, the partial pressure of these residual gases is reduced.

In a furnace constructed of molybdenum or tungsten, which when heated will not react with the residual gases in the system, the reduction of the partial pressure of the oxidizers within that particular heating chamber can be effected only by a further reduction of the total pressure in the system.

Operating pressures of 0.1 microns Hg (0.0133 Pa) (10^{-4} Torr) (diffusion pump range) are customary for general heat treating in a furnace with metallic radiation shields. In a pure graphite lined furnace, pressures of 50 to 100 microns, which can be attained easily with only a mechanical pump, may be used.

38.00 EFFECTS OF FURNACE ATMOSPHERES ON REFRACTORIES

The selection of the proper refractory material is an important consideration in sintering furnaces that have the refractory lining exposed to the furnace atmosphere. Various types of furnace atmospheres have different effects on refractories, and conversely, the refractories have an effect on the furnace atmosphere. If the proper refractory is not selected, very high maintenance costs may result. If extremely low dew points are required, as in the case of sintering stainless steel, the selection of a refractory that will not react with dry hydrogen to form water vapor is very important if the furnace is of the non-muffle type.

38.11 Carbon Monoxide Atmosphere

Both endothermic and exothermic reducing atmospheres contain carbon monoxide. Carbon monoxide is stable at elevated sintering temperatures, but under some conditions breaks down into carbon and carbon dioxide in the temperature range between 800 to 1200F (425 to 650C). When an atmosphere gas containing carbon monoxide is free of carbon dioxide and low in water content it is unstable and needs only the promoting effect afforded by a catalytic surface to cause a rapid reaction to take place.

The equilibrium of the reaction may be represented in the following equation:

$$2\,CO = C + CO_2 \qquad \textit{Eq. 38-1}$$

Iron oxide and certain other oxides such as chromium and nickel act as a catalyst for this reaction. Therefore, in selecting refractories that are exposed to carbon monoxide, it is important to select as low an iron-oxide content as practical.

The iron-oxide particles act as nuclei to start the reversible reaction. Carbon begins to deposit in globules, as shown in Figure 38-1. The globules then keep growing until the carbon ruptures and disintegrates the refractory brick. This reaction is, of course, more critical in the zones where the temperature may be between 800 to 1200F (425 to 650C), as in burn-off, purge and insulated cooling zones. However, the refractory in the high-temperature sintering section is also subjected to this reaction because at some point below the hot face of the refractory there will be a zone between 800 to 1200F (420 to -650C). This effect is shown in a light-weight insulating brick in Figure 38-2.

The same reaction, of course, occurs in hard fire brick. In hard fire brick another reaction, besides carbon deposition, seems to occur which breaks down the refractory bond and the refractory disintegrates into dust. This is shown in Figure 38-3.

Some furnace manufacturers have done considerable research on this subject, and have set up quality control on the refractory used in atmosphere furnaces. In the extremely critical "burn-off" and / or pre-heat chambers, and

FIGURE 38-1
Carbon globules in light weight refractory.

in insulated cooling chambers, extra precaution should also be taken to use high temperature refractory, 2800F to 3000F (1540 to 1650C), instead of the less expensive lower temperature grades, 1500 to 1600F (815 to 871C). The reason is that any residual iron oxide in the brick would combine with the silica, alumina and clays at the high firing temperature the refractory was subjected to in manufacture, and thus, the free iron oxide will not be present to act as a catalyst. This, of course, leads to a more expensive furnace, but in the long run will pay many times over the initial cost in the savings through low furnace maintenance. The same rule applies to hard fire brick. Only super-duty fire brick should be used in furnaces containing a carbon monoxide atmosphere, regardless of the temperature of the operation.

Deposition of carbon is harmful to the refractory and to heating elements mounted on the refractory face. The elements may short out between turns or between the lead-in terminals. Since carbon is an excellent conductor, the voltage should be kept as low as practical on controlled atmosphere furnaces.

Even with the proper refractory, the furnace must be operated and maintained properly to assure good performance and low maintenance. When carbon monoxide-type atmospheres are used, the furnace must be burned

FIGURE 38-2
Carbon core in light weight refractory.

FIGURE 38-3
Carbon globules and disintegration of refractory bond.

out periodically to remove any carbon that may have deposited on the brick work or heating elements. On continuous furnaces, this should be done at least once a week, and more often if possible. This is especially important in furnaces where there may be a danger of carry-over of the lubricants from the burn-off zone. If excessive carbon builds up in the front of the sintering zone, steps should be taken to reduce the amount of lubricant used in the compact, to raise the temperature of the burnoff chamber or add a separate burn-off chamber to the front of the furnace, or to improve the counterflow of atmosphere gas

38.12 Hydrogen Atmosphere

Under some furnace conditions, hydrogen will reduce the oxides of iron, silicon, and various other metals in most refractories except high-purity alumina. However, pure-alumina refractories can be reduced in a hydrogen atmosphere having an extremely low dew point, such as -140F (-96C), and a furnace temperature of 2400F (1315C). The higher the temperature, and the lower the dew point, the more severe the reaction and the greater the number of oxides reduced. This type of reaction will cause the bond to disintegrate, and the refractory will crumble. Where the refractory is exposed to high-hydrogen dry atmospheres in sintering furnaces, extreme care must be taken to choose the proper grade of refractory to withstand the temperature of operation.

38.13 Atmospheres Reaction on Dew Point

When oxides are reduced, water vapor will be formed and the dew point of the atmosphere in the furnace will rise, regardless of the entering dew point. In molybdenum furnaces, where it is impossible to use alloy muffles because of the high operating temperature, the choice of the refractory for the hydrogen atmosphere is extremely important when low dew points are required. The refractory must be of the highest alumina available. Any silicon oxide or soda (NaO) in the refractory may be reduced to silicon or sodium and volatilized, then reoxidized in the case of silicon, or form sodium hydroxide in the case of sodium. Both will condense in the cooler parts of the furnace in the form of a white powder or glass wool. Care must be taken that silica and soda be excluded from the binder of the alumina refractory, as well as from the cement used for laying up the refractory.

38.14 Atmosphere Conductivity and Refractory Heat Loss

Hydrogen has approximately five times the thermal conductivity of air. Therefore, expect higher heat losses for an equivalent thickness of insulation when hydrogen-rich atmospheres are used. When designing a furnace of tight-shell-construction, this must be considered in calculating heat losses. Refractory and insulation manufacturers' data usually give the heat loss in terms of an air atmosphere. When high-hydrogen reducing atmospheres are used, a greater thickness of insulation must be added to the furnace wall, or the shell will run hotter with greater heat loss.

The relative thermal conductivities for the various reducing atmospheres in reference to air are: hydrogen 5 times as great, dissociated ammonia 3.75 times as great, endothermic 2 times as great, and rich exothermic only slightly greater. Therefore, it is necessary to base the heat-loss calculations on the particular atmosphere to be used. The refractory manufacturer should be consulted on heat loss data for a specific atmosphere.

PART FOUR

OPTIONAL OPERATIONS

40.00 OPTIONAL OPERATIONS

Many designers welcome powder metallurgy as a manufacturing process because functional parts can be made completely by a simple pressing and sintering operation. However, some applications may require additional treatment subsequent to sintering. For some secondary operations P/M parts can be handled the same as their wrought equivalents, but in some cases the P/M part may have to be treated differently. When it is known how these differences affect the operation, then post treatments become routine.

Some of the more important of these subsequent treatments are repressing, machining, heat treating, plating, tumbling, infiltrating, impregnating, and joining.

41.00 REPRESSING

The terms repressing, sizing and coining refer to an operation in which a sintered part is returned to a die and again subjected to pressure. Repressing may be used to reshape the sintered part, densify it or to control critical size tolerances. The pressures used may be equal to or greater than the original briquetting pressure. Ductility is usually decreased as a result of the cold work imparted during repressing, and the repressed parts will have lower impact strength. To improve this, parts are sometimes resintered after repressing. Parts may be repressed either in the same die used for briquetting or in a second set of tools. The choice here is influenced by the dimensional change which occurs during the initial sintering operation. Usually a lead-in or chamfer is placed in the die to allow the part to enter the die cavity more easily. Prior to repressing the parts must be lubricated by the application of zinc stearate, molybdenum disulfide or some other lubricant.

Parts can be repressed on compacting presses, modified presses, or presses specifically designed for repressing. In general, repressing rates are faster than compacting rates, being limited only by the rate at which parts can be fed to and removed from the die and by allowable temperature build-up in the die due to friction. The coining or sizing press feeding mechanism must be capable of high speeds as well as extremely accurate positioning of the part. A popular restrike feeder is an indexing dial type. This feeder is usually mechanically driven by the press motion and brought into final position by an alignment pin or a precision cam box. A restrike press with a high speed dial feeder is shown in Figure 41-1. It is important that safeguards be incorporated in the automatic feeder to ensure that there is only one part in the die for each stroke of the press. Another type of press is illustrated in Figure 41-2. This machine uses a removable die set and can be used for coining, sizing, or for compacting Class I and Class II parts. The transfer mechanism which is connected and driven mechanically by the press mechanism is a four-station, walking-beam, linear system.

FIGURE 41-1

Restrike press with dial feeder.

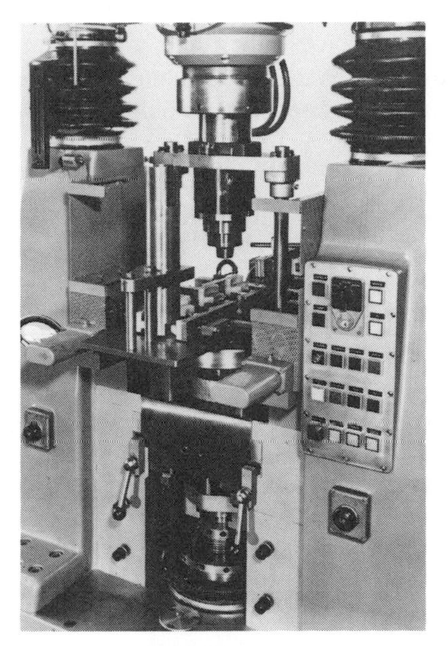

FIGURE 41-2
Coining press

41.11 Sizing

Sizing is used to refine dimensional accuracy, or to compensate for warpage or other defects which may occur during sintering. The sized parts will be straighter, dimensional tolerances will be closer, and surface finish will be improved. Generally, little or no increase in density is achieved since pressures used are usually no more than the initial compacting pressure. Bearings are usually sized to achieve better dimensional control, but the sizing operation must be done so that maximum surface porosity is maintained.

41.12 Coining

Coining not only improves dimensional accuracy but the density of the part is increased by the use of higher pressures. Because the sintering operation has soft annealed the part, considerable plastic flow is possible during coining. The increase in mechanical properties is in many cases so significant that a soft, unalloyed P/M part gains sufficient strength for use under quite severe conditions.

Coining is also used to engrave or emboss designs or identifying marks on one or both faces of the part.

42.00 MACHINING

Although P/M parts are often specified because machining operations can be eliminated, in some cases it is more economical to leave certain part details designed for machining, rather than to incorporate them in the pressed configuration. In this respect, machining is probably the most important of the mechanical finishing operations. Usually the form of the finished part is compacted with repetitive accuracy on hundreds of thousands of parts. Thus, design of chucks, jigs, mandrels, and fixtures for machining is simplified. Normally, if the porosity of the part is less than 10%, machining of P/M parts is similar to machining wrought or cast parts. However, machining problems increase with increasing part porosity due to interrupted cuts, shifting loads and shock and vibration. Annular grooves, threads, undercuts, holes at right angles to the pressing direction, are examples of some forms that do require secondary machining. When machining P/M parts, both the basic material and machining techniques must be considered.

42.11 Iron Parts

These are the most difficult parts to machine because they are soft. The soft iron tends to tear easily resulting in a poor surface finish. Straight iron must be machined slowly, using feed rates of less than .004 in (0.1 mm) per revolution with a cut 0.02 to 0.03 in (0.5 to 0.75 mm) deep. The tools are prone to cratering type wear. The machinability of straight iron parts can be improved if they are repressed before machining which will give them added hardness due to a cold working effect. As mentioned, density also influences machinability in that less tool wear occurs at higher densities.

42.12 Iron-Copper Parts

The presence of copper increases the machinability of iron base parts because of its hardening effect. For example, with 3% copper in the mix, the feed rate can be increased to approximately twice that for straight iron and still obtain good surface finishes with no tearing. With still higher copper contents, even better surface finishes can be obtained under the same machining conditions.

Iron parts infiltrated with copper are the easiest to machine. Many machinists consider this P/M material better to cut than a free machining steel.

42.13 Iron-Carbon Parts

The addition of carbon to the mix causes the formation of pearlite during sintering and a subsequent hardening of the matrix. However, carbon is not as beneficial toward increasing machinability as is copper. For example, a part containing 0.3% combined carbon can be machined at feed rates about 1-1/2 times that of straight iron, but only about 3/4 the rate of a 3% copper part.

In general, it can be said that almost any additions to an iron mix which will have a hardening effect on the matrix will improve machinability. Copper in large amounts improves the surface finish and also gives longer tool life. Carbon by itself gives only a slight improvement. In many mixes both carbon and copper are present and the combined effects of these elements will be noticed. Other elements that may be added to the powder to give better machinability are sulfur, lead, and phosphorus. All of these result in a smoother surface finish on the machined part. However, sulfur in the form of iron sulfide will improve tool life. If manganese is present in the base material, manganese sulfide will be formed which will act as a chip breaker and give a smoother surface finish, but will tend to decrease tool life. Phosphorus has the general effect of decreasing tool life, although smoother surface finishes can be obtained at faster machining rates. Lead has an embrittling effect on iron.

Recent development of atomized high-compressibility steel powders has extended the range of P/M parts that can be machined. Several low carbon mild steels are made to the chemistry of free machining wrought steels. In addition, they all have high machinability ratings. For example, on a numberical rating system in which the more difficult the material is to machine the lower the number, an iron-carbon (0.60% combined) is rated at 5, a copper-infiltrated iron at 20 and a 1% copper 1% nickel, 0.8% carbon resulfurized steel was 10. The higher sulfur contents permit machining with heavier feeds at higher speeds, with increased tool life, better surface finishes and closer tolerances. The resulfurized steels should be used: when parts require a large amount of machining; when material cost is secondary to machining cost; when set-up time is short compared to the length of the run; when inspection costs are high due to closer tolerances, better surfaces, etc.

42.14 Stainless Steel Parts

The machining of sintered stainless steel parts is no more difficult than machining wrought stainless steel parts. The parts fabricator must take extreme care to assure that the sintered structure is truly austenitic and free of carbide precipitation at the grain boundaries. Poor machinability of stainless steel can be attributed to carbide precipitation resulting from improper sintering. The use of a spray mist lubricant during drilling, tapping or any other machining operation is definitely recommended. For machining stainless steel, the use of carbide tools is almost mandatory. Most fabricators use 303L, an austenitic stainless steel in which sulfur has been alloyed.

42.16 Aluminum

Because of the excellent compressibility of aluminum, 90% of theoretical density at 12 to 15 tsi (165 to 206 N/mm^2) and 95% theoretical at 25 tsi (345 N/mm^2), sintered aluminum parts are high in density. Consequently, most machining operations can be performed without difficulty. Any problems that occur are normally the result of improper lubricant removal before sintering or excessive aluminum oxide content. The excessive oxide is caused by improper temperature or dew point control during sintering.

42.17 Bronze

When it is necessary to machine a bronze sleeve bearing, the main object is to maintain a good surface finish without closing the surface porosity. The surface porosity must be kept open so that the oil present in the pores can get to the surface to act as a lubricant.

As a rule, it is better to size the ID of a bearing rather than to machine it. However, in those cases where it is absolutely necessary to machine the ID of a bearing, it should be bored with a single point tool and not drilled or reamed. Rough boring can be done without too much concern about surface finish; however, when the final 0.005 to 0.010 in (0.13 to 0.25 mm) are reached, the techniques must be modified to obtain the best open pore surface. Finished boring should be thought of as a fine threading operation in that a sharp pointed tool should be used which will cut cleanly and not smear the surface.

TABLE 42-I
Bronze Finishing Tool

	Material - Grade C-4 Carbide
(a) Back Rake Angle	35°
(b) Side Relief	10°
(c) End Relief	10°
(d) Side Cutting Angle	35°
(e) End Cutting Angle	50°
Sharp point with no nose radius	

The cutting speed should be between 100 to 150 sfm (31 to 46 m/min) to maintain maximum porosity. The depth of the finish cut should not exceed 0.002 in (0.05 mm). The use of fine feeds in the order of 0.001 to 0.002 in (0.025 to 0.05 mm) per revolution will give clean cut surfaces.

It is difficult to mill, drill or thread sintered bronze and maintain surface porosity. Grinding is definitely not recommended because of the extreme tendency to close the surface pores and charging the pores with abrasive particles.

43.10 Machining Practices for P/M Parts

Although the speeds and feeds discussed here encompass general ranges for machining P/M parts, many parts producers do a bit of experimenting before setting up final machining parameters. If some doubt exists, it is common practice to refer to standard machining charts for cast or wrought parts of the same composition and then increase the recommended speeds 10%. Ultimately, the penalty for obtaining better surfaces (not smeared) is lower tool life. The use of lubricants or coolants during machining is definitely recommended if structural parts are being machined and porosity serves no functional purpose. However, it is sometimes necessary to machine dry parts in which the coolant would contaminate the oil in self-lubricating bearings or when the parts must be painted, plated or heat treated after machining. A jet of air at the cutting point for cooling and chip removal has proved successful in machining porous P/M parts.

43.11 Turning and Boring

For these operations, it has been found that carbide tools are best. The tool should be designed with a sharp nose point. Any tool dressing should be examined at magnifications of 25 to 40 diameters for "saw toothing."

TABLE 42-II
General Turning Tool - Single Point

Material - Grade C-4 Carbide	
Side Cutting Angle	15°
End Cutting Angle	15°
Back Rake	10°
End Relief	10°
Side Relief	10°
1/32 in (0.8 mm) nose radius for roughing cuts	
Sharp point for finishing cuts	

Cutting speeds of 175 to 350 sfm (53 to 107 m/min) are satisfactory. It has been found that the maximum depth of cut should be in the order of 0.005 in. (0.127 mm) and that fine feeds of 0.001 to 0.003 in. (0.025 to 0.075 mm) per revolution give the best results. For roughing cuts, and where a porous surface is not required, speeds can be increased to 500 sfm (152 m/min).

43.12 Drilling

The drilling of an oil hole in a sintered bushing is a common example of machining P/M parts. The use of carbide or HSS drills with a low right hand helix angle will prevent the drill from digging in. Speeds of 70 sfm (21 m/min) for high speed steel and up to 200 sfm (61 m/min) for carbide drills are used. It is always best to use a mechanical feed on the drill in order to get optimum performance. For a 1/8 in (3 mm) diameter hole, a feed rate of 0.002 in (0.05 mm) per revolution has been found to be satisfactory, whereas, for a 1 in (25.4 mm) diameter hole, .010 in (0.254 mm) per revolution can be used. The rates for intermediate sizes are proportional.

43.13 Tapping

Tapping is easily accomplished by using conventional carbide or HSS taps. In most cases, we are not concerned with porosity when tapping a part. It is recommended that two-fluted taps be used for holes less than 5/16 in (8 mm) in diameter. Three-fluted taps can be used for holes greater than 5/16 in (8 mm) in diameter.

Spiral pointed taps are most desirable because the spiral point throws the chips ahead and prevents them from driving into the pores. If difficulties are encountered, the relief of the tap can be increased to nearly twice that used for conventional ferrous materials.

43.14 Milling

Milling is generally difficult because of the tendency of soft sintered materials to smear. It is recommended that helical tool cutters with an axial rake be used so that the chips are sheared on an angle to minimize the tendency to smear. The milling cutter must be dead sharp. For high speed steels, speeds should be about 70 sfm (21 m/min), whereas, carbide cutters can be run as high as 300 sfm (91 m/min). For roughing cuts, the feeds should be about 0.10-0.15 in (2.5 to 3.8 mm) per tooth, whereas, finishing feeds should be finer and in the order of 0.002 to 0.005 in (0.05 to .13 mm) per tooth. Slots and undercuts are examples of milling in P/M parts.

43.15 Shaping

Shaping is, of course, very similar to turning or boring in that a single point tool is used. The recommendations for turning and boring should be followed. However, it is important that the operator makes sure that the tool does not drag on the return stroke which will tend to mar the finish. In general, a good surface finish cannot be obtained during shaping since the available feeds on most machines are usually too coarse.

43.16 Reaming

Reaming of bearings or bushings is not recommended if porosity is to be maintained because of the tendency to smear the bearing surface. However, structural parts are often reamed. For small holes up to 3/8 in (10 mm), the drill should leave a reaming allowance of about 0.002 in (0.05 mm) on the diameter. For 1 in (25 mm) holes, this can be increased to 0.005 in (0.13 mm). Reamers should be used in floating holders and run at 25 to 50 sfm (8-15 m/min). Feed rates of the same magnitude as are used for drilling are recommended.

43.17 Ball Sizing or Burnishing

In general, ball sizing or burnishing has been found best for maintaining the porosity in the holes on sintered copper or bronze bearings. No more than 0.002 in/in (0.05 mm/mm) of diameter should be displaced during the burnishing operation. In general, the smallest amount of material that can be cleaned up successfully is preferred because surface porosity is lost in proportion to the amount of material displaced. Most any of the conventional burnishing tools of the button-type or roller-type will produce satisfactory finishes. It is best to do ball sizing on the bearing after it is assembled into the housing. For absolute control of the ID, the springback of the material must be considered. This close-in is in the order of 0.002 to 0.003 in/in (0.05 to 0.075 mm/mm) of diameter and is influenced by wall thickness of the bearing and burnishing allowance.

43.18 Broaching

Broaching is not advised if porosity is to be held because of the tendency to smear the metal. Broaching of structural parts is possible and if this machining method is used, it is recommended that at least .010 in (0.25 mm) of stock should be removed. Broaches with angular cutting edges are recommended.

43.19 Grinding, Honing and Lapping

These operations are definitely not recommended for porous parts. Structural parts can be finished in this way using hard, fine-grain wheels at light feed and conventional speeds. It is recommended that a light-bodied oil be used to flush away the abrasive particles to obtain the best surface finish.

We have discussed P/M materials and various types of machining. We must consider the surface finish, the speed of material removal and the wear on the tool. There is little argument that carbide tools are generally best for machining P/M parts. However, carbide tools are not mandatory for all operations, and in some cases their use cannot be justified on a cost basis. For example, since most P/M parts are small, carbide tools are prone to breakage, especially if dull. They cost about 15 times more than HSS and rarely offer equivalent tool life in small size ranges below 1/8 in (3.2 mm) for drilling and 1/4 in (6 mm) for tapping. On the other hand, larger carbide drills and taps pay for themselves quickly.

44.00 HEAT TREATMENT

Ferrous P/M parts may require some heat treatment after sintering. This treatment may be through hardening, case carburizing, or carbonitriding. Generally, heat treatment is thought of in connection with carburizing and hardening but we will also consider annealing, steam treatment and precipitation hardening.

Heat treating processes for P/M parts are generally the same as for wrought parts, but additional factors of porosity and chemistry must be considered.

44.11 Iron Parts

Straight iron parts can be case hardened or, if they are first through-carburized, they will respond to hardening.

44.12 Carbon Steel Parts

These parts have had the carbon added as an ingredient in the powder mix and as such will respond to a hardening heat treatment. Normally, they are not case hardened.

44.13 Copper Steels

Copper does not increase hardenability but decreases the solubility of carbon in the ferrite. Parts made of these materials are commonly hardened by a simple quenching operation.

44.14 Nickel Steels

This class of material responds well to heat treatment which results in improved impact strength. Carbon control is critical in these compositions.

44.15 Stainless Steels

Only the martensitic stainless steels (400 series) can be hardened by heat treatment. The austenitic steels (300 series) will not respond to a quenching heat treatment.

44.16 Non-Ferrous Materials

In general, these cannot be hardened except by cold working. There are some exceptions in age hardenable compositions but these are seldom used in powder metallurgy.

44.17 Alloy Steels

These materials are made from alloy powders or mixtures of elemental powders and respond to deep hardening as long as their compositions are homogeneous. These include the 4600, 4300, 4400 and 8600 grades currently available.

44.20 Heat Treating Practices

44.21 Through Hardening

It is sometimes necessary to obtain better wear resistance or to increase the strength or impact resistance of a P/M part. One of the lest expensive ways to increase strength requirements is to heat treat or harden the metal powder part by quench hardening or through hardening.

Basically, the through hardening process involves heating the sintered metal powder part above the critical temperature (A_3) (normally about 1500F (815C), then quenching it to convert the solid (austenite) to the hard constituent martensite. This process increases strength and hardness but reduces ductility. During the heating process, the metal powder part must be protected from oxidation by surrounding the part with a neutral or slightly carburizing atmosphere. Parts are usually quenched in well circulated oil. Quenching porous parts in water or brine introduces the problem of internal rusting.

Physical properties obtained are dependent on the carbon content of the part to be heat treated, the density of the sintered compact and carbon potential in the hardening furnace atmosphere. If the compact is made from carbon-free iron, the amount of carbon introduced into the part by the furnace atmosphere determines the properties obtained. Highest strength values of parts will occur at a combined carbon content of .87%. Lower carbon contents will give lower tensile strengths and less wear resistance, whereas, higher combined carbons tend to diminish the strength and ductility.

The quenching medium is important because it directly affects the extent of hardening. Since the rate of cooling in oil is slower than in water, the resulting hardness obtained using an oil quench may be much less than by using water. To get around this, some parts producers add ammonia to the furnace atmosphere. The ammonia serves to retard the rate of transformation, thus giving better hardness and strength. The amount of ammonia added should be limited to 2% as excessive amounts may cause the soft austenite to remain in the structure after quenching.

Quench oil temperature should be held about 275F (135C) to keep it free of water. A clean standard quench oil should be well circulated around the parts during the quench operation. In dealing with P/M parts it should be remembered that the structure is porous and, therefore, quenched parts will contain from 10 to 25% by volume of oil within the porous structure. Any attempt to charge quenched parts back into the furnace could be dangerous. The oil will be forced out of the pores by the heat and may cause a serious fire.

The quench oil may be removed by soaking the part in a chlorinated solvent for at least four hours. After soaking, an air dry for twenty four (24) hours should allow the solvents to evaporate.

44.22 Case Hardening

Where a hard, wear-resistant skin is desired, case hardening is used. This process is usually accomplished by adding carbon to a sintered part which had little or no carbon in the original mix.

Methods of performing this treatment on parts are pack carburizing, cyaniding and gas carburizing, and induction heating. In pack carburizing, parts are packed in carburizing compounds such as hardwood charcoal or coke and subjected to heat. CO decomposes into C and CO_2) and the carbon is absorbed at the part surface. Diffusion also passes this carbon on to the interior of the part at the carburizing temperature of about 1700F (925C). The process is generally used where deep cases are desirable. Control of the carbon entering the part cannot be controlled, hence the case is rather high in carbon at the surface.

Cyaniding is seldom used for P/M parts because of the difficulty encountered in removing the cyanide from the parts and the toxic nature of the cyanide compounds.

Gas carburizing is probably the most universally accepted method of carburizing ferrous P/M parts, particularly those containing no copper. The carbonitriding process is a modification of the straight gas carburizing process in that anhydrous ammonia is added to the atmosphere of the furnace along with the hydrocarbon gas and endothermic gas. At the carburizing temperature, the ammonia dissociates into hydrogen and nitrogen with some of the nitrogen combining with the iron and the hydrogen being given off to the furnace atmosphere. The higher the carburizing temperature, the less nitrogen will be taken into the part. The nitrogen not only forms a hard wear resistant compound with iron, but speeds the formation of carbon with iron and retards the rate of transformation. In general, where light case depths of extremely hard surfaces are desired, carbonitriding is the most desirable practice.

By definition, carburizing and carbonitriding case hardening augment the surface carbon while keeping the interior of the part soft and more ductile. If the density of the powder part is below 7.2 g/cm³, there is enough interconnected porosity so that the carburizing gases will go into the pores and make the interior of the part hard, thus defeating the purpose of case carburizing. At 7.2 g/cm³ and above, the pores are minimized so that practically all the carbon is taken into the part by starting at the surface and diffusing inward. A clearly defined case is obtained with the higher densities, whereas, little or no line of demarcation is visible at the lower densities.

The depth of case is a function of time and temperature. An example of case depths which may be expected at a temperature of 1500F (815C) with a 1% carbon potential on a 7.2 g/cm³ part would be as follows:

TABLE 44-I
Carburizing Time vs. Case Depth

Time	Depth
5 Min. at Temperature	0.0025 in 0.06 mm
10 Min. at Temperature	0.0037 in 0.09 mm
15 Min. at Temperature	0.0045 in 0.11 mm
20 Min. at Temperature	0.0053 in 0.13 mm
25 Min. at Temperature	0.0058 in 0.15 mm
30 Min. at Temperature	0.006 in 0.16 mm

The rate of diffusion of carbon and nitrogen in P/M parts is much faster than for wrought products because of their porosity. Temperatures are generally reduced therefore so that depth of case may be controlled more closely. Temperatures between 1450 to 1500F (790 to 815C) seem to provide the best uniformity of case structure.

Induction hardening is used on sprockets, cams and gears provided sufficient combined carbon is present in either iron or iron-copper parts. The parts are passed through an induction coil for 1-5 seconds and oil quenched. Temperatures up to 10% higher than those for through hardening are used.

The very short time at the higher temperatures is possible since little grain growth occurs. Gear teeth can be hardened with little effect on tolerances of inside diameters.

44.23 Annealing or Stress-Relieving

Annealing of metal powder parts is generally performed to reduce cold working stresses introduced in a repressing operation or to alter the structure for improved machinability. From the standpoint of relieving working stresses, a sub-critical anneal at 1150F (620C) will be helpful and will not affect dimensions seriously. In cases where the sintering process is such that drastic cooling gives a rather hard structure, annealing at 1375F (745C) followed by a slow cool may give more desirable machining structures.

Atmosphere control during annealing is necessary and it has been found that the annealing of powder parts may take somewhat longer than wrought parts due to the slower rate of heat transfer in the medium density materials.

44.24 Precipitation Hardening

Parts made from iron and copper, or iron, carbon, copper alloys, or parts infiltrated with copper materials can be increased in strength and ductility by solution treating and precipitation hardening. At the sintering temperature, the gamma iron will hold about 8.5% copper in solution, whereas, just after transformation takes place only 2% Cu can be held by the alpha iron. As the temperature drops, copper is thrown out of solution to the point where only .04% Cu can be held at room temperature. This process of expelling copper provides the hardening process.

Parts are heated to 1540F (840C), then water quenched. Reheating the parts to 930F (500C) for two hours will allow copper to precipitate in the grain boundaries. Tensile strengths as high as 65,000 psi (450 N/mm^2) are possible with an 85% iron-15% copper alloy. Elongations of 7.0% at 80 R_B apparent hardness were obtained.

It is significant that tensile strengths can be raised as much as 25,000 psi (170 N/mm^2) by the precipitation hardening process without the presence of carbon in the compact.

44.25 Steam Treating

Because of the porous nature of medium density P/M parts, the steam oxidizing process can be used to advantage to provide increased surface wear and corrosion resistance. The object of the process is to coat all exposed surfaces, interior and exterior, with a hard black magnetic iron oxide. The flat, black finish enhances wear and gives low-reflection surfaces which have a more pleasing color. As outside surfaces are worn away, new oxide surfaces are exposed which provide hard wearing surfaces. It is known as "the poor man's heat treatment," although companies far from poor use it. It is extensively used in the firearms and photocopy industries.

As a general rule steam oxidized parts are immersed in oil to help both corrosion and wear characteristics. Where the part is protected from outdoor weather, an oiled and steam oxidized part may provide service for many years without showing signs of corrosion.

In general, a steam oxidized part is confined to applications requiring low tensile stress and high compressive yield strength along with good wear and corrosion resistance. Tensile strength, impact resistance and ductility are lowered by the process.

Parts to be steam treated are placed in a wire mesh blanket or tray, which in turn is placed in a forced convection furnace. Parts must be free from oil or grease. Parts are first heated to 700F (370C) to drive off moisture. Steam is then introduced into the furnace to purge the air from the furnace. The purging process may take 10 or 15 minutes for 2 ft^3 (0.56 m^3) of parts. The temperature is raised to 950-1000F (510-540C). At these elevated temperatures, the iron in the parts combines with the oxygen in the steam to form Fe_3O_4. The hydrogen from water is given off to the furnace atmosphere. As with any hydrogen atmosphere, care must be exercised to prevent an explosion. If the volume of steam is too low at the reaction temperature, a reddish-brown soft oxide coating will result.

It should also be noted that if steam comes into contact with the parts before the temperature is up to 212F (100C), the parts will rust.

The coating forms rapidly on the parts. The degree of oxide formation may be determined by the increase in weight of the part. The table indicates the % weight increase as the reaction takes place.

TABLE 44-II
Weight Increase From Steam Oxidizing

	30 Minutes
Density g/cm^3	% Weight Increase
5.7	6.1
5.8	5.0
5.9	4.3
6.0	3.2
6.2	2.4
6.4	1.9

After the first 30 minutes, the amount of iron exposed for conversion to oxide becomes quite small and further processing is of little value.

When steam oxidizing P/M parts, the voids are partially filled with oxide which increases their density and hardness. The volume available for the storage of oil decreases. It will be observed that time of the oxidizing operation must be controlled to prevent complete closing of the voids and, therefore, elimination of the pores. Hardness values obtained before and after steam oxidizing are as follows:

TABLE 44-III

Before R_B - 0 (Apparent)	After R_B - 100 (Apparent)
K_{500} - 285	K_{500} - 340

44.30 Furnaces for P/M Heat Treatment

44.31 Carburizing and Carbonitriding

Iron and steel P/M parts can be heat treated after sintering, similar to parts made of wrought metals. The strength and hardness of medium and high-carbon steel parts can be increased by heating to 1500-1650F (815-900C) (depending on the carbon content) and quenching in oil. Low-carbon steel parts can be case-hardened by carbonitriding or carburizing, followed by an oil quench. Atmosphere methods for case-hardening are almost universally applied, because salt has a tendency to be absorbed into the pores of the parts, lowering their corrosion resistance. The best results of case-hardening are obtained on P/M parts having a density of 7.4 g/cc or higher. With a density of 7.4 g/cc, a well-defined martensitic case similar to that produced on wrought steel can be obtained.

The atmosphere used for case-hardening is composed of the endothermic type previously described, adjusted to a high-carbon potential of 20F (-6.7C) dew point, with 5 to 10% natural gas added to increase the carburizing potential. If it is desired to obtain a case similar to that produced by cyanide, about 5% natural gas with 10% ammonia is added to the endothermic carrier gas. This treatment is referred to as carbonitriding. If the P/M part is composed of medium- or high-carbon powders, the dew point of the endothermic atmosphere is adjusted to the carbon content of the part, as shown in the equilibrium diagram (Fig. 36-1).

The batch type, vertical gas fired radiant tube furnace shown in Figure 44-1 is ideal for neutral hardening, carbonitriding, or straight gas carburizing. However, this furnace can also be electrically heated, with a low voltage

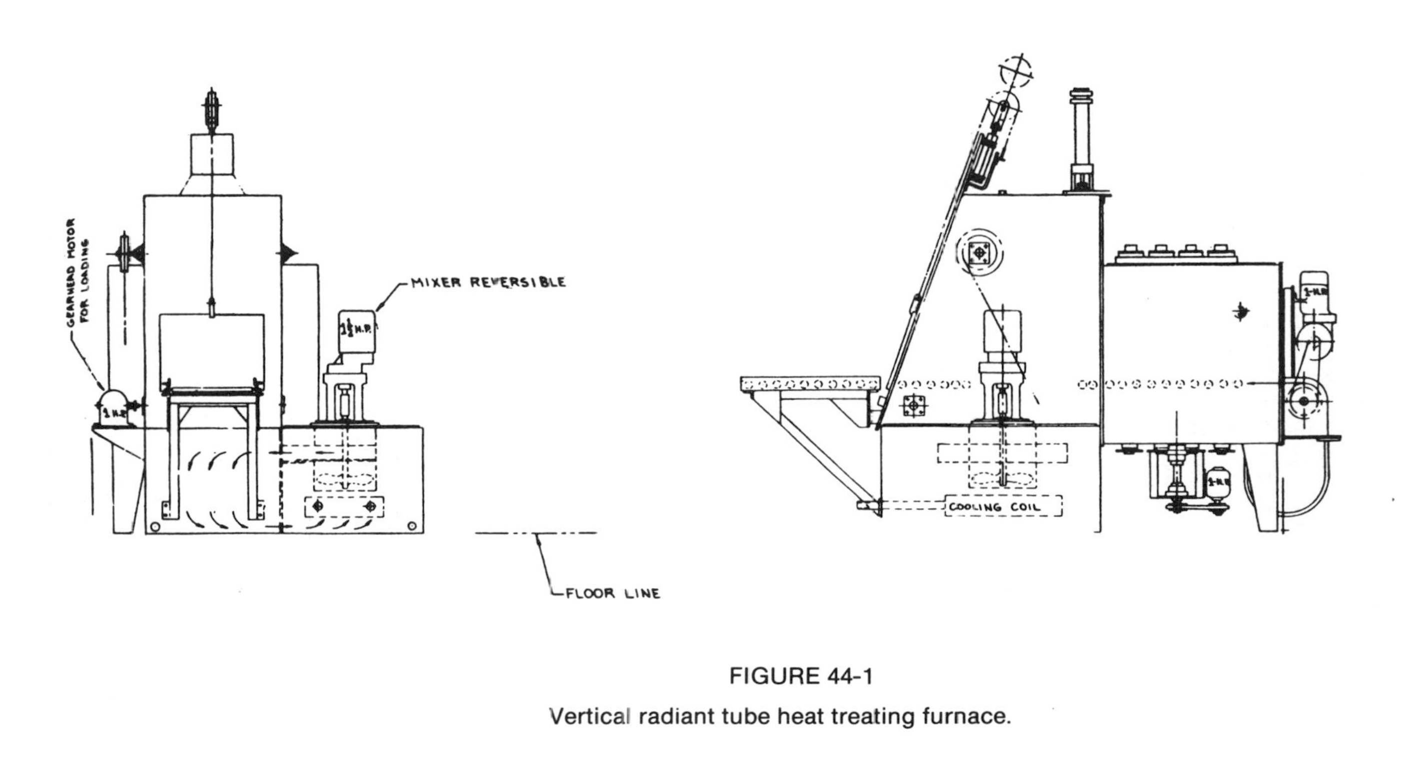

FIGURE 44-1
Vertical radiant tube heat treating furnace.

resistor developed for carburizing and carbonitriding atmospheres. This furnace can be either completely automatic, or hand operated. The work is charged into the purge and quench vestibule by hand. In the automatic furnace, the work is carried into the furnace by a mechanical pusher. At the end of the heating cycle, the work is discharged automatically from the furnace onto the quench elevator, and then quenched. After quenching for the proper time, as set on the automatic timers, the work is raised automatically, and a signal notifies the operator to pull the tray out of the purge chamber. While one tray of work is in the quench tank, another is charged automatically into the furnace. Since the work does not contact air until completely heated and quenched, the parts stay bright and free from scale throughout the entire cycle.

Where production does not warrant automatic operation, or where it is desired to keep the initial cost low, the furnace can be operated manually. The above furnace has the capacity of neutral hardening up to 400 lbs/hr (180 kg/hr), or light case carburizing or carbonitriding up to 300 lbs/hr (136 kg/hr).

44.32 Steam-Oxidizing or Blueing

After iron parts have been sintered, it is sometimes desirable to impart a steam-oxidized surface or blueing effect. This is done to certain iron parts for wear resistance, color, or corrosion resistance. To obtain the best corrosion resistance, the parts should be tumbled in an oil-impregnated sawdust or quenched into soluble oil.

A steam tempering furnace is shown in Figure 44-2. The furnace usually is of the electric 100% forced-convection type constructed gas-tight to hold a steam atmosphere. In operation, the parts to be treated are loaded in the 500F (260C) furnace with the steam turned off. After the parts are preheated, the steam is turned on. The furnace is allowed to purge for about 30 minutes to remove the air. The temperature is raised to 1000F (540C) and the load allowed to soak approximately 30 minutes or more. The load of parts can then be quenched into soluble oil for a deep blue-black color where oil impregnation is desired, or allowed to drop to 800F (425C) and then taken out to cool in air. The finish of the air-cooled parts is not as dark in color as the oil-treated parts.

44.33 Vacuum Heat Treating

Neutral hardening and carburizing of P/M parts may be accommodated in a partial pressure environment.

Carburizing in a partial pressure is accomplished by introducing a hydro-

FIGURE 44-2
Steam tempering furnace.

carbon gas (natural gas, CH_4) into a heated vacuum furnace.

This gas produces a carburizing atmosphere that varies in strength from zero to the saturation point. Work processed under these conditions exhibits an absorption of carbon characteristic to its behavior at a specific temperature. The potential of the atmosphere is a function of the partial pressure of the carbon and not the total pressure of the system. When the carburizing gas is admitted, carburization is instantaneous. Absorption of carbon immediately ceases upon the evacuation of the carburizing gas. The rate of carbon absorption increases with the increase of partial pressure of carbon. The maximum rate of carbon absorption occurs at the saturation point of carbon vapor at a specific temperature. For instance, low carbon steel heated to 1900F (1038C) and subjected to a saturated carbon will swiftly absorb carbon to a saturated surface carbon concentration.

A typical partial pressure neutral hardening and carburizing furnace is shown in Figure 44-3.

44.40 Hardness Testing

Because of the porosity of most P/M parts, conventional Rockwell hardness readings actually reflect a composite hardness resulting from the porous structure. Because sliding and surface contact involves contact between individual particles, the heat treated P/M material's microhardness is a more accurate indication of its scoring resistance.

The proper determination of case depths is a study in itself. Years ago, case depth quality was checked by sliding a test file across the surface. If the file failed to bite into the surface, the hardness was considered satisfactory. Depth of case was determined by fracturing the part and comparing the brittle fractured case with various sizes of shim stock. The demand for higher quality has introduced more precise methods to evaluate quality and depth of case. Parts are now sectioned so that the case can be observed microscopically at right angles to the carburized surface.

A Knoop microhardness indentor is used to check the hardness of the case. The Rockwell hardness tests have been found wanting for evaluation of case because of the large indentor sizes involved. It should also be remembered that in Rockwell testing, the support of the part has an important bearing on the accuracy of the test. In a microhardness test, a prepared specimen is checked by pushing a diamond with a known load on it into a particle. The depression is measured and hardness calculated.

45.00 IMPREGNATION

The porosity of P/M parts can be controlled for impregnation with oil for self-lubricating qualities. However, porosity which is detrimental to plating or machining of P/M parts can be filled with a resin.

45.11 Oil Impregnation

The ability to provide self-lubrication to bearings and even in selected areas of a part is unique to P/M. The type and viscosity of oil used must, of course, be compatible with the application.

Parts to be impregnated may be simply submerged in an oil bath for several hours. Reasonably good penetration can be obtained if the oil temperature is maintained at about 180F (80C). The best results are obtained with vacuum impregnation. In this treatment, the parts to be impregnated with oil are placed in a basket in a vacuum chamber. After the air in the

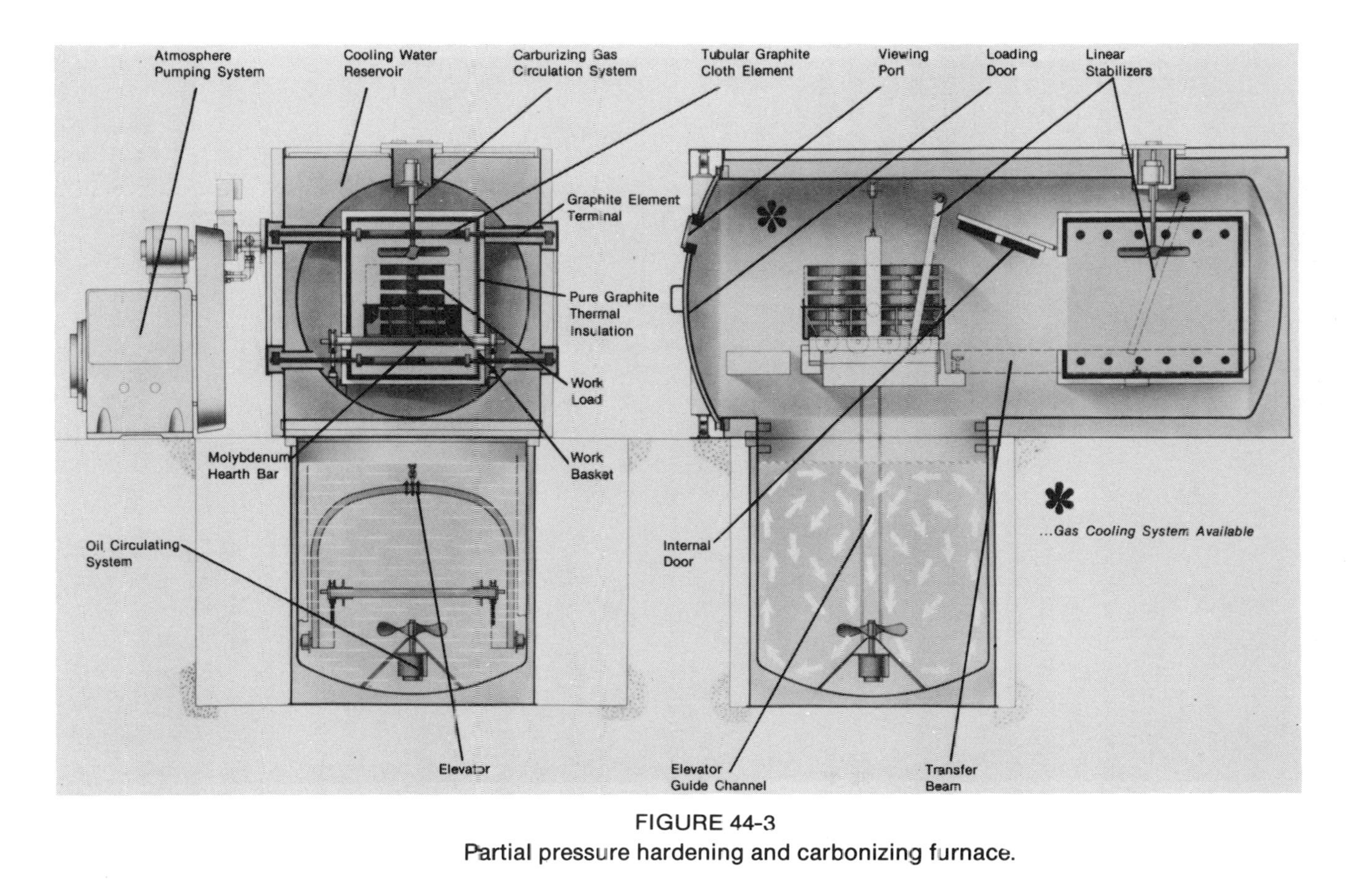

FIGURE 44-3
Partial pressure hardening and carbonizing furnace.

chamber, and the parts, has been evacuated, the chamber is flooded with oil. Next, the chamber is returned to atmospheric pressure, the oil drained from the chamber and the surfaces of the parts.

45.12 Resin Impregnation

Impregnation with a plastic such as a polyester resin greatly improves the machinability of P/M parts by filling the pores and increasing the density. On the average, a plastic impregnated part of medium density has up to 500% better machinability than the same unimpregnated part. One obvious drawback is that machined plastic impregnated parts cannot be heat treated.

Porosity is undesirable also in a part to be plated because of entrapment of plating or cleaning solutions in the pore structure. Porous outside surfaces also make it difficult to obtain a smoothly plated surface. Impregnation with plastic eliminates both these problems. This can be accomplished by soaking the parts in the molten plastic. Vacuum impregnation gives faster and better results. De-aeration of the pores is followed by the pores gulping up the impregnant when the vacuum is released.

46.00 INFILTRATION

For structural engineering components, where high strength and toughness are important, it is generally desirable to eliminate residual porosity in the P/M part. The mechanical properties of porous metal are impaired not only by the absence of metal in the pores, but also by the stress-raising action of the pores themselves.

A porous metallic body having a substantial degree of interconnected porosity, i.e., the skeleton, is infiltrated with another metal of lower melting point. This is carried out above the melting temperature of the infiltrant, but below the melting point of the skeleton. The infiltrant is completely absorbed as a liquid into the pores of the skeleton by capillary action to produce a component with a composite structure, which may be virtually solid. Such components exhibit physical and mechanical properties generally comparable with those of solid metal.

Copper is used widely as an infiltrant material for ferrous based materials. It is used where higher strengths and pressure tightness are required, for example, in hydraulic pumps and valves.

Plastic and resin infiltrant materials are also used. They are generally less expensive than copper, they improve machinability and are commonly used for closing porosity on components that are to be plated for corrosion protection.

47.00 SURFACE PROTECTION

47.11 Electroplating

P/M parts having less than 10% porosity are electroplated by standard procedures. To plate parts having more than 10% porosity, it is important to first impregnate the pores with a resin or close them by peening the surface to exclude the plating salts from the pores of the part.

47.12 Thermal Passivation

This is a technique used for the protection of stainless steel P/M parts. It consists simply of heating parts for 30 minutes in air at 620 to 930F (325 to 500C) during which time a passive film develops. This film improves the resistance of the parts to certain corrosive attacks. Its value is in passivating the surfaces of the part to the bottom of the pores. This improves the resistance of the parts to pitting corrosion.

47.13 Steam Treatment

This is accomplished by heating parts to 1000F (540C) and subjecting them to superheated steam under pressure. (See also Sections 44.25 and 44.32.)

47.14 Cementation

Articles to be coated are heated in a rotating furnace together with powdered coating metal.

47.15 Coatings for Aluminum

Aluminum is difficult to electroplate because it has a high affinity for oxygen.

Using chemical conversion coatings of chromates, phosphates and aluminum oxide, gold, green or gray colors can be obtained.

Anodizing is done using either chromic acid or sulfuric acid. Hard coat anodizing uses sulfuric-oxalic acid. Following cleaning and before sealing, single-source colors of organic or inorganic materials can be applied.

47.20 Plating Practices

47.21 Copper Plating

Acid copper plating seems to give the best results on metal powder parts because of the excellent throwing power of the process and the tolerance of the bath for impurities. Because copper tends to oxidize or tarnish, very little plating is done with copper alone. Where the copper color is desirable, a lacquer or coating is put over the bright copper immediately after plating and drying. Most plating is used as an undercoating for nickel or for nickel and chromium.

A typical acid copper plating process is as follows:

Copper Sulfate	- 23%
Sulfuric Acid	- 4%
Temperature	- 100F agitated bath (38C)
Current Density	- 50 amps/ft² (540 A/m²)
Voltage	- 6 volts

Deposits of 0.002 in. (0.05 mm) may be obtained in 45 minutes and the deposit may be buffed to smooth luster for nickel or chromium plating. This acid copper deposit is somewhat softer than that from cyanide copper baths.

In multiple plating applications, copper is used as a leveler and sealer and may be put on in deposits ranging up to 0.005 in. (0.13 mm). The nickel deposit, next, provides the support, hardness and corrosion resistance for a subsequent wear resistant chromium. The nickel deposit is from 0.005 to 0.01 in (0.13 to .25 mm) thick, and the chromium deposit is from 1×10^{-5} to 3×10^{-5} in (2.5×10^{-4} to 7.5×10^{-4} mm).

47.22 Nickel Plating

Nickel plating is used either alone or with copper undercoat to enhance wear resistance and to provide a bright corrosion resistant surface. Nickel is used on steel, brass and other base metals. It is the oldest protective/decorative electro-deposited metallic coating.

	Watts Bath	Sulfamate	Fluborate
Ni Sulfate %	23-43	0-3	0-1.8
Ni Chloride %	3-6	27-47	23-31
Ni Sulfamate %		27-47	
Ni Fluborate %			23-31
Boric Acid %	3-5	3-5	1.5-3
Ph	1.5-5.2	3-5	2.5-4
Temperature F (C)	115-160	100-140	100-160
	(46-71)	(38-60)	38-71)
Current Density, A/ft²	10-100	25-300	25-300
A/m²	108-1080	270-3240	270-3240
Hardness, VHN	100-250	130-160	125-300
Elongation, %	10-35	3-30	5-30

47.23 Chromium Plating

There are many types of chromium plating; however, we can generally classify Cr as decorative or hard chrome. The decorative Cr is generally 1×10^{-5} to 2×10^{-5} in (2.5×20^{-4} to 5×10^{-4} mm), .01 to .02 mils in thickness, and is put on a part generally for corrosion protection and appearance. Hard chrome is usually a heavier deposit and is plated for the purpose of wear and for heat resistance. Decorative chrome, like metal powder parts, generally contains pores. We will confine our discussion to decorative chrome plating.

Decorative chrome is generally put on over a more ductile and very corrosion resistant nickel plate on many wrought products. This practice may not be necessary for metal powder parts which are to be oil impregnated. Where the part has a supply of oil, a film will be present over the entire chrome plated part, so excellent corrosion resistance is obtained.

The most generally used chromium plating bath is as follows:

Chromic Acid	- 33 oz/gallon (25.8%)
Sulfate Ratio	- 100/1
Temperature	- 115F (46C)
Current Density (Cathode)	- 150 A/ft² (1620 A/m²)

47.24 Zinc and Cadmium Plating

Cadmium has greater throwing power and will go into depressions in a plated surface more readily than will zinc. Cadmium plates faster, hence, may cut down on the labor involved. However, cadmium is about 13 times greater in cost than zinc. Zinc will withstand greater concentrated salt spray than cadmium; however, at 80 feet (25 m) from the ocean cadmium and zinc are comparable in resistance to corrosion. Cadmium is usually used indoors in radio and electronic type instruments because the corrosion products formed are of lower volume than the products formed on zinc plated parts. Cadmium has a lower coefficient of friction with steel than does zinc and is used where soldering is involved to promote better bonding. Cadmium is never used around food processing equipment because of the possible toxic results.

Although the volume of these metals plated on powder parts is small, the methods used are as follows:

Zinc Plating Water Solution

	Zinc Cyanide	6.25%
	Na Cyanide	3.91%
	NaOH	9.38%
Temperature		Room Temperature
Current Density		30 A/ft² (324 A/m²)
Rate of Deposit		.0002 in (0.005 mm) in 6 min.

Cadmium Plating Water Solution

	Cadmium Oxide	2.73%
	Cadmium Metal	2.39%
	Sodium Cyanide	11.95%
	Sodium Carbonate	3.91%
Temperature		Room Temperature
Current Density		30 A/ft² (324 A/m²)
Rate of Deposit		.0002 in (0.05 mm) in 4 min.

Control of both the zinc plating bath and the cadmium bath is much the same as for plating wrought products. The bath may be contaminated at a faster rate when powder parts are plated in volume.

48.00 JOINING

48.11 Joining During Sintering

This method of joining achieves the equivalent of a "shrink" fit by using two materials with different growth characteristics during sintering.

The gear cluster shown in Figure 48-1 is made of parts a and b. The larger gear, a, is made from a grade of iron powder which has only 0.1% growth during sintering. The smaller gear, b, has an inner boss, and is made from a mixture of iron and copper powders which give 0.6% growth in sintering.

The green compacts are produced to give a clearance of 0.002 to 0.003 in (0.05 to 0.08 mm) between the bore of a part a and the boss on part b. The parts are assembled before sintering and are permanently joined during sintering both by the growth of part b into part a, and also by the tendency of the copper in part b to diffuse into part a.

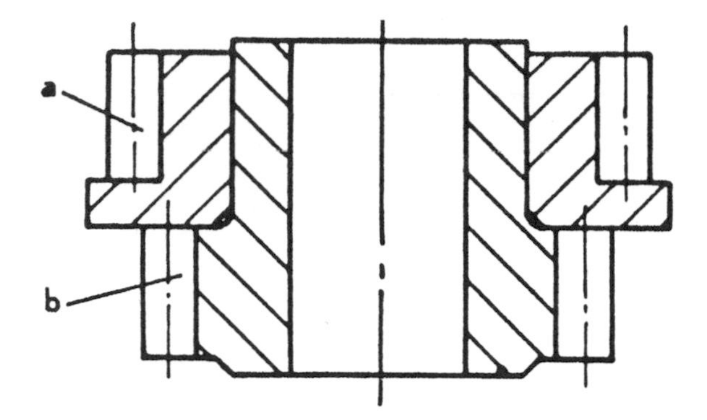

FIGURE 48-1
Joining P/M parts during sintering.

By an inversion of the above method, the shrinkage characteristics of a material can be used to close it permanently upon another part having zero shrinkage. In cases where heavy loading will occur or the alignment of teeth between a and b is important, keys, flats or splines can be formed on the joining faces as the part is being pressed.

48.12 Joining During Infiltration

Parts can be joined during the infiltration process by assembling the component parts and infiltrating the assembly during sintering. Once the porosity within the parts is completely filled with the infiltrant, a bond similar to furnace brazing is achieved.

48.13 Projection Welding

Sintered parts are reliably joined to sintered parts, or to conventionally machined parts, by using projection welding.

The method is illustrated by Figure 48-2 which shows a gear with a cam and a cylindrical part with a clutch.

A ring shaped projection and several small projections, which are formed during compacting provide the metal necessary for welding. During the welding operation the two parts are kept aligned by an insulated shaft through their respective central holes.

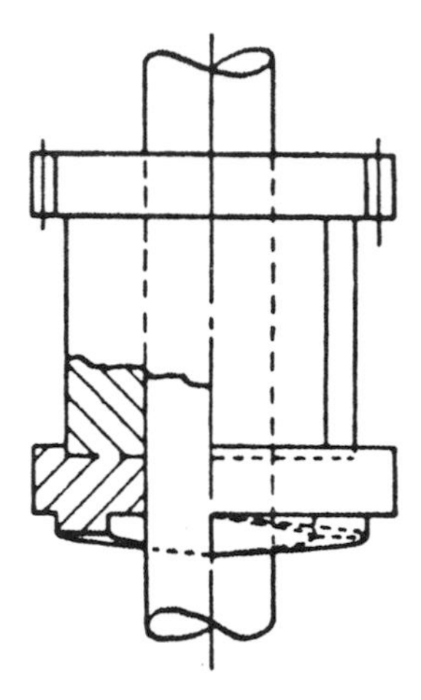

FIGURE 48-2
Assembling P/M parts by projection welding.

48.14 Joining by Threaded Fasteners

Two sintered parts can be joined, or a sintered part can be joined to a machine-made part, by the conventional method of using threaded fasteners. Figure 48-3 shows a ratchet and gear unit made in two pieces. The ratchet teeth are case hardened to resist wear. The gear teeth do not require case hardening, as satisfactory physical characteristics in this case can be obtained by sintering and sizing only. The bearing surface in the neck between gear and ratchet is also preferred unhardened.

In the design shown, the two-piece assembly permits the case hardening of the ratchet alone. The drive between ratchet and gear is provided by the slightly tapered projection at the joining face, and three screws with recessed heads secure the pieces together on assembly.

The shapes of the two parts are suitable for molding and sizing. The only additional operations necessary are threading of the three screw holes and case hardening the ratchet.

49.00 TUMBLING

P/M parts can be deburred, burnished or polished using conventional barrel, vibratory or centrifugal finishing machines. Normally, plastic, stone or ceramic media such as aluminum oxide or silicon carbide are used dry or in slurries of water or water-soluble compounds.

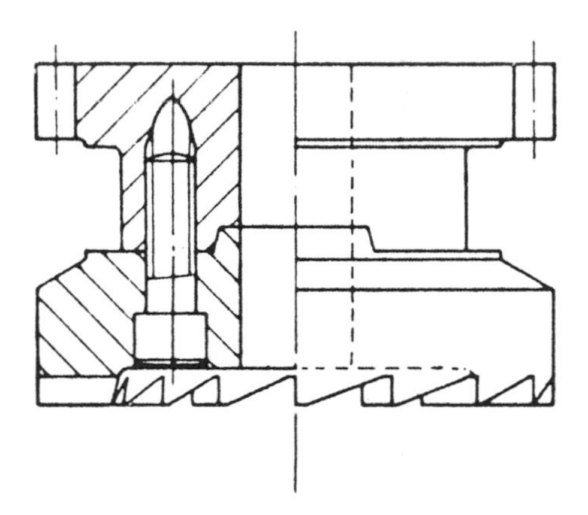

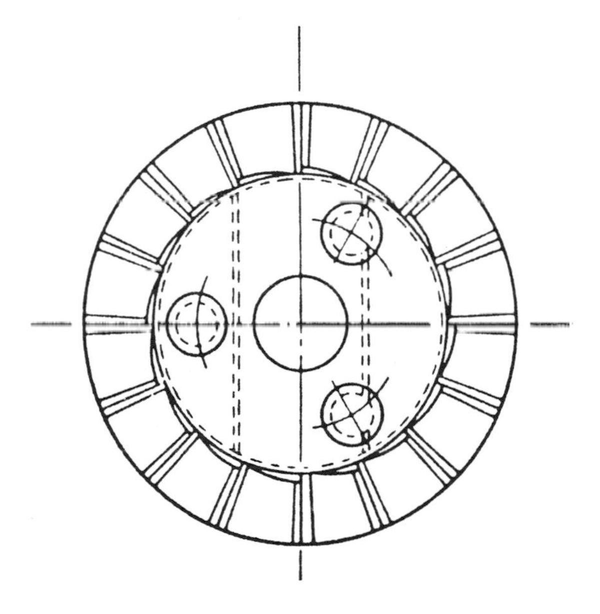

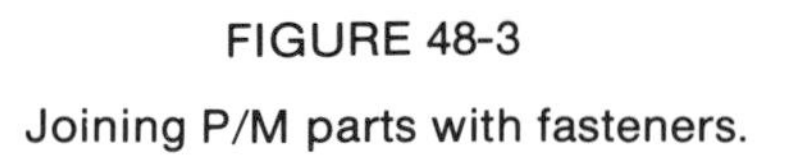

FIGURE 48-3

Joining P/M parts with fasteners.

For barrel finishing, it is important that the barrel volume be about 1/2 full and that rotational speed be adjusted to give a sliding action to deburr the work. By changing the media, barrel size, compound, rotation speed, etc., a large degree of flexibility can be gained. Normally, a 3 to 1 media/parts ratio is used. There is virtually no minimum size of parts which can be successfully barrel finished. However, only 20 to 30% of the total load is actually worked on at any moment. Maximum size of parts is limited only by available equipment. A good rule to follow is that the major dimension of free-tumbling parts be less than one-third the diameter of the barrel.

Whereas, barrel machines must operate in batches, continuous flow can be achieved in vibratory equipment. It is more effective for cutting down, radiusing, and surface scoring; and is used to a lesser extent for cleaning, polishing, burnishing or coloring. Vibrators are normally run at 90 to 95% of volume with a 3 to 1 media/parts charge. Even at reduced speeds and amplitudes, vibratory finishing is much faster than barrel finishing.

Unless P/M parts are impregnated prior to finishing, fine particles of abrasive and liquid will enter the pores. These abrasives and compounds should be removed by rinsing, spin drying, and heating in an oven to vaporize any water. A rust inhibitor should then be used on all ferrous parts.

If compatible with the production cycle, deburring is best done while the metal is in a strain-hardened condition such as after coining or sizing. In this state the extruded metal particles are very brittle and shear off readily.

PART FIVE

GLOSSARY

51.00 TERMS FOR METAL POWDER PRESSES AND TOOLING MPIF Standard #35

This standard defines and illustrates schematically those segments of a metal powder compacting press and the tooling used therein which are directly involved in forming a powder metallurgy compact. Its purpose is to provide the manufacturer and user of such equipment with a uniform and universally acceptable glossary of terms.

Compacting Tool Set—the part or parts making up the confining form in which a powder is pressed. Parts of the tool set may be some or all of the following: die, punches, and core rods. (Synonymous with mold.)

Upper Punch Holder—that part of the press which is fastened to the upper ram.

Upper Punch Adapters—a sleeve or a member which may be required to adapt the upper punch to the upper punch holder.

Upper Punch—member of the tool set that closes the die and forms the top of the part being produced.

Die—member of the tool set that forms the periphery of the part being produced.

Die Liner—an insert sometimes used inside of a die body, to prolong die life.

Die Body—a member which retains or supports the die liner.

Die Table—the opening in the press for retaining the die.

Die Adapter—a sleeve or component which may be required to adapt the die to the die table bore.

Die Table Bore—the bore in the die table to fit the outside diameter of the die or its adapter.

Die Table Counterbore—the bore in the die table with tapped holes used for clamping the die in place.

Die Table Clearance Hole—the hole in the bore of the die table necessary to permit clearance for the lower punch when two counterbores are used.

Lower Punch—a member of the tool set that determines the powder fill and forms the bottom of the part being produced. Secondary or sub-divided lower punches may be necessary to facilitate filling, forming and ejecting of multiple level parts.

Lower Punch Holder—that part of the press which is fastened to the lower ram.

Lower Punch Adapter—a sleeve or component which may be required to adapt the lower punch to the lower punch holder.

Core Rod—member of the die set that forms a through hole in the compact.

Core Rod Support—the press member supporting the core rod. It may be movable or stationary.

Core Rod Adapter—the press member used to adapt the core rod to the core rod support.

52.00 TERMS USED IN POWDER METALLURGY MPIF Standard #09

Absolute Pore Size — The maximum pore opening of a porous material, such as a filter through which no larger particle will pass. Synonymous with maximum Pore Size.

Acicular Powder — Needle shaped particles.

Activated Sintering — A sintering process during which the rate of sintering is increased, for example by addition of a substance to the powder or by changing sintering conditions.

Agglomerate — Several particles adhering together.

Air Classification — The separation of powder into particle size fractions by means of an air stream of controlled velocity.

Alloy Powder — See Pre-alloy Powder.

Angle of Repose — The basal angle of a pile formed by a powder when freely poured under specified conditions onto a horizontal surface.

Apparent Density — The weight of a unit volume of powder, usually expressed as grams per cubic centimeter, determined by a specified method.

Apparent Hardness — The value obtained by testing a sintered material with standard indentation hardness equipment. Since the reading is a composite of pores and solid material, it is usually lower than that of solid material of the same composition and condition. Not to be confused with particle hardness.

Atomization — The dispersion of a molten metal into particles by a rapidly moving gas or liquid stream or by mechanical means.

Atomized Metal Powder — Metal powder produced by the dispersion of a molten metal by a rapidly moving gas or liquid stream, or by mechanical means.

Binder — A cementing medium; either a material added to the powder to increase the green strength of the compact, and which is expelled during sintering; or a material (usually of relatively lower melting point) added to a powder mixture for the specific purpose of cementing together powder particles which alone would not sinter into a strong body.

Blank — A pressed, presintered or fully sintered compact, usually in the unfinished condition, requiring cutting, machining, or some other operation to give it its final shape. See Preforming.

Blending — The thorough intermingling of powders of the same nominal composition (not to be confused with mixing).

Bridging — The formation of arched cavities in a powder mass.

Briquet — See Compact.

Bulk Density — The density of a powder under non-specified conditions, for example in a shipping container.

Burn Off — That stage of a sintering cycle referring to the time and temperature necessary to remove ingredients used to assist the forming of a powder metallurgy part, such as binders or die lubricants.

Cake — A coalesced mass of unpressed metal powder.

Carbonyl Powder — A metal powder prepared by the thermal decomposition of a metal carbonyl.

Cemented Carbides — A solid and coherent mass made by pressing and sintering a mixture of powders of one or more metallic carbides and a much smaller amount of a metal, such as cobalt, to serve as a binder.

Chemically Precipitated Metal Powder — Powder produced by the replacement of one metal from a solution of its salts by the addition of another element higher in the electrochemical series, or by other reducing agent.

Classification — Separation of a powder into fractions according to particle size.

Closed Pore — A pore not communicating with the surface.

Coining — The pressing of a sintered compact to obtain a definite surface configuration (not to be confused with Repressing or Sizing).

Cold Pressing — The forming of a compact at room temperature.

Cold Welding — Cohesion between two surfaces of metal, generally under the influence of externally applied pressure at room temperature.

Comminuted Powder — A powder produced by mechanical disintegration of solid metal.

Communicating Pores — See Interconnected Porosity.

Compact — An object produced by the compression of metal powder, generally while confined in a die, with or without the inclusion of nonmetallic constituents. Synonymous with Briquet.

Compactibility — See Compressibility.

Compacting Tool Set — See Die.

Composite Compact — A metal powder compact consisting of two or more adhering layers, rings, or other shapes of different metals or alloys with each layer retaining its original identity.

Composite Powder — A powder in which each particle consists of two or more separate materials.

Compound Compact — A metal powder compact consisting of mixed metals, the particles of which are joined by pressing or sintering or both, with each metal particle retaining substantially its original composition.

Compressibility — The capacity of a metal powder to be compacted uniaxially in a closed die. Compressibility may be expressed numerically as the pressure to reach a required density or, alternately, the density at a given pressure. Synonymous with Compactibility.

Compression Ratio — The ratio of the volume of the loose powder to the volume of the compact made from it. Synonymous with Fill Ratio.

Continuous Sintering — Presintering, or sintering, in such manner that the objects are advanced through the furnace at a fixed rate by manual or mechanical means.

Core Rod — The separate member of the compacting tool set or die that forms a hole in the compact.

Cracked Ammonia — See Dissociated Ammonia.

Cut — See Fraction.

Dendritic Powder — Particles, usually of electrolytic origin, having the typical pine tree structure.

Density (Dry) — The weight per unit volume of an unimpregnated P/M part.

Density (Wet) — The weight per unit volume of a P/M part impregnated with oil or other nonmetallic materials.

Density Ratio — The ratio of the determined density of a compact to the absolute density of metal of the same composition, usually expressed as a percentage. Synonymous with % Theoretical Density.

Die — The part or parts making up the confining form in which a powder is pressed. The parts of the die may be some or all of the following: Die body, punches, and core rods. Synonymous with Mold and Compacting Tool Set.

Die Body — The stationary or fixed part of a die.

Die Insert — A removable liner or part of a die body or punch. Synonymous with Die Liner.

Die Lubricant — A lubricant mixed with the powder or applied to the walls of the die and punches to facilitate the pressing and ejection of the compact.

Die Set — The parts of a press that hold and locate the die in proper relation to the punches.

Dimensional Change — See Growth; See Shrinkage.

Dispersion-Strengthened Material — A material consisting of a metal and a finely dispersed, substantially insoluble, metallic or nonmetallic phase.

Dissociated Ammonia — A reducing gas produced by the thermal decomposition of anhydrous ammonia over a catalyst, resulting in a gas of 75% hydrogen and 25% nitrogen. Synonymous with Cracked Ammonia.

Double-Action Pressing — A method by which a powder is pressed between opposing punches both moving relative to the die.

Electrolytic Powder — Powder produced by electrolytic deposition or the pulverization of an electrodeposit.

Endothermic Atmosphere (Gas) — A reducing gas atmosphere used in sintering and produced by the reaction of a hydrocarbon fuel gas and air over a catalyst with the aid of an external heat source. It is low in carbon dioxide and water vapor with relatively large percentages of hydrogen and carbon monoxide. Maximum combustibles approximately 60%.

Equi-axed Powder — See Granular Powder.

Exothermic Atmosphere (Gas) — A reducing gas atmosphere used in sintering and produced by partial or complete combustion of a hydrocarbon fuel gas and air. Maximum combustibles approximately 25%.

Exudation — The action by which all or a portion of the low melting constituent of a compact is forced to the surface during sintering. Sometimes referred to as "bleed out". Synonymous with Sweating.

Ferrite — In the field of magnetics, substances having the general formula: $\mathbf{M^{+}O, M_2^{+++}O_3}$ the trivalent metal often being iron.

Fill Ratio — See Compression Ratio.

Fines — The portion of a powder composed of particles which are smaller than a specified size, currently less than 44 micrometres. See also Superfines.

Flake Powder — Flat or scale-like particles whose thickness is small compared with the other dimensions.

Flow Rate — The time required for a powder sample of standard weight to flow through an orifice in a standard instrument according to a specified procedure.

Fluid Permeability — See Permeability (Fluid).

Fraction — That portion of a powder sample which lies between two stated particle sizes. Synonymous with Cut.

Gas Classification — The separation of powder into particle size fractions by means of a gas stream of controlled velocity.

Granular Powder — Particles having approximately equi-dimensional non-spherical shapes.

Granulation — A general term for the production of coarse metal powders using methods such as:

a. Pouring the molten metal through a screen into water (shotting),

b. Violent agitation of the molten metal while solidifying,

c. Agglomeration of smaller particles.

Green — Unsintered (not sintered); for example: Green com-pact, green density, green strength.

Green Expansion — The increase in dimensions of a compact relative to the die dimensions after being ejected from die, measured at right angles to the direction of pressing.

Growth — An increase in dimensions of a compact occurring during sintering (converse of Shrinkage).

Hot Densification — The rapid deformation of a heated powder preform in a die set, primarily in the pressing direction, in order to reduce porosity. Synonymous with Hot Repressing.

Hot Forging — The rapid deformation of a heated powder preform in a die set, involving appreciable lateral movement, in order to reduce porosity.

Hot Pressing — The simultaneous heating and molding of a compact.

Hot Repressing — See Hot Densification.

Hydrogen Loss — The loss in weight of metal powder or of a compact caused by heating a representative sample for a specified time and temperature in a purified hydrogen atmosphere—broadly a measure of the oxygen content of the sample, when applied to materials containing only such oxides as are reducible with hydrogen and no hydride forming material.

Hydrogen Reduced Powder — Powder produced by the hydrogen reduction of a metal oxide.

Impregnation — A process of filling the pores of a sintered compact with a nonmetallic material such as oil, wax or resin.

Infiltration — A process of filling the pores of a sintered, or unsintered, P/M compact with a metal or alloy of lower melting point.

Infiltration Efficiency — The ratio in percent of the amount of infiltrant absorbed by the part to the amount of infiltrant originally used. See Standard 49.

Infiltration Erosion — The solution of the base metal at the contact area between the part and the molten infiltrant where the infiltrant flows into the part causing pits, channels and coarse surface porosity. See Standard 49.

Infiltrant Loading Density — Infiltrant weight per area of contact between infiltrant and part. See Standard 49.

Infiltration Residue — Material that remains on the surface of the part after infiltration. See Standard 49.

Interconnected Porosity — A network of contiguous pores in and extending to the surface of a sintered compact. Usually applied to P/M materials where the interconnected porosity is determined by impregnating the specimens with oil.

Irregular Powder — Powder having particles lacking symmetry.

Isostatic Pressing — The pressing of a powder by subjecting it to nominally equal pressure from every direction.

K - Factor — The strength constant in the formula for "Radial Crushing Strength" of a plain sleeve specimen of sintered metal. See Radial Crushing Strength.

Lamination — A rupture in the pressed compact caused by the mass slippage of a part of the compact. Synonymous with Pressing Crack and Slip Crack.

Liquid Phase Sintering — Sintering of a P/M compact, or loose powder aggregate under conditions where a liquid phase, is present during part of the sintering cycle.

Lower Punch — The member of the compacting tool set or die that determines the volume of powder fill and forms the bottom of the part being produced. Secondary or sub-divided lower punches may be necessary to facilitate filling, forming and ejecting of multiple level parts.

Lubricant — An agent mixed with or incorporated in a powder to facilitate the pressing and ejecting of the compact.

Magnetic Permeability — See Permeability (Magnetic).

Master Alloy Powder — A pre-alloyed powder of high concentration of alloy content designed to be diluted when blended with a base powder to produce the desired composition.

Matrix Metal — The continuous phase of a polyphase alloy or mechanical mixture; the physically continuous metallic constituent in which separate particles of another constituent are embedded.

Maximum Pore Size — See Absolute Pore Size.

Mechanical Component — A shaped body pressed from metal powder and sintered wherein self-lubrication is not a primary property. A part but not an oil- impregnated bearing. See Powder Metallurgy Part.

Mesh — The screen number of the finest screen through which substantially all of the particles of a given sample will pass. The number of screen openings per linear inch of screen.

Metal Filter — A metal structure having controlled inter-connected porosity produced to meet filtration or permeability requirements.

Metal Powder — Discrete particles of elemental metals or alloys normally within the size range of 0.1 to 1000 micrometres.

Milling — The mechanical treatment of metal powder, or metal powder mixtures, as in a ball mill, to alter the size or shape of the individual particles, or to coat one component of the mixture with another.

Minus Sieve — The portion of a powder sample which passes through a standard sieve of specified number. See Plus Sieve.

Mixed Powder — A powder made by mixing two or more powders as uniformly as possible. The constituent powders will differ in chemical composition and/or in particle size and/or shape.

Mixing — The thorough intermingling of powders of two or or more materials.

Mold — See Die.

Molding — The pressing of powder to form a compact.

Multiple Pressing — A method of pressing whereby two or more compacts are produced simultaneously in separate die cavities.

Neck Formation — The development of a necklike bond between particles during sintering.

Needles — Elongated rod-like particles.

Nodular Powder — Irregular particles having knotted, rounded, or similar shapes.

Oil Content — The measured amount of oil contained in an oil-impregnated object, for example a self-lubricating bearing.

Open Pore — A pore communicating with the surface.

Oversize Powder — Particles coarser than the maximum permitted by a given particle size specification.

Packing Material — Any material in which compacts are embedded during the presintering or sintering operation.

Partially Alloyed Powder — See semi-alloyed powder.

Particle Size — The controlling lineal dimension of an individual particle as determined by analysis with sieves or other suitable means.

Particle Size Distribution — The percentage by weight, or by number of each fraction into which a powder sample has been classified with respect to sieve number or micrometres. (Preferred usage: "Particle size distribution by weight" or "particle size distribution by frequency.")

Percent Theoretical Density — See Density Ratio.

Permeability (Fluid) — The rate of fluid flow through a porous material under specified conditions of area, thickness and pressure.

Permeability (Magnetic) — A measure of the ease of a material to be magnetized, not to be confused with fluid permeability.

Platelet Powder — Flat particles of metal powder having considerable thickness (as compared to flake powder).

Plus Sieve — The portion of a powder sample retained on a standard sieve of specified number. See Minus Sieve.

P/M — The acronym representing powder metallurgy. Used as P/M Part. P/M Product, P/M Process, etc.

P/M — A shaped object that has been formed from metal powders and bonded by heating below the melting point of the major constituent. A structural or mechanical component, made by the powder metallurgy process.

Pore — An inherent or induced cavity within a particle or within an object.

Pore-Forming Material — A substance included in a powder mixture which volatizes during sintering and thereby produces a desired kind and degree of porosity in the finished compact.

Porosity — The amount of pores (voids) expressed as a percentage of the total volume of the powder metallurgy part.

Pore Size — The average pore diameter of a porous material, such as a filter, which conforms to specific particle removal requirements usually the removal of 95% to 100% of a given particle size distribution.

Porous Metal — A metal structure having controlled inter-connected porosity. See Metal Filter.

Powder Flow Meter — An instrument for measuring the rate of flow of a powder according to a specified procedure.

Powder Metallurgy — The arts of producing metal powders and of the utilization of metal powders for the production of massive materials and shaped objects.

Powder Metallurgy Part — See P/M Part.

Powder Rolling — The progressive compacting of metal powders by the use of rolling mill. Synonymous with Roll Compacting.

Pre-Alloyed Powder — A metallic powder composed of two or more elements which are alloyed in the powder manufacturing process, and in which the particles are of the same nominal composition throughout. Synonymous with Alloy-Powder.

Preforming — The initial pressing of a metal powder to form a compact which is subjected to a subsequent pressing operation other than coining or sizing. Also, the preliminary shaping of a refractory metal compact after presintering and before the final sintering.

Premix — A uniform mixture of ingredients to a prescribed analysis, prepared by the powder producer, for direct use in compacting powder metallurgy products.

Presintering — The heating of a compact at a temperature below the normal final sintering temperature, usually to increase the ease of handling or shaping the compact, or to remove a lubricant or binder prior to sintering.

Pressed Bar — A compact in the form of a bar; a green compact.

Pressed Density — The weight per unit volume of an unsintered compact. Synonymous with Green Density.

Pressing Crack — See Lamination.

Pulverization — The reduction in size of metal powder by mechanical means.

Punch — Part of a die or compacting tool set which is used to transmit pressure to the powder in the die cavity. See Upper Punch and Lower Punch.

Radial Crushing Strength — The relative capacity, of a plain sleeve specimen made by powder metallurgy, to resist fracture induced by a force applied between flat parallel plates in a direction perpendicular to the axis of the specimen.

Rate-of-Oil-Flow — The rate at which a specified oil will pass through a sintered porous compact under specified test conditions.

Reduced Metal Powder — Metal Powder produced, without melting, by the chemical reduction of metal oxides or other compounds.

Repressing — The application of pressure to a previously pressed and sintered compact, usually for the purpose of improving physical property.

Roll Compacting — See Powder Rolling.

Rolled Compact — A compact made by passing metal powder continuously through a rolling mill so as to form relatively long sheets of pressed material.

Rotary Press — A machine fitted with a rotating table carrying multiple dies in which a material is pressed.

Screen Analysis — See Sieve Analysis.

Segment Die — A die made of parts which can be separated for the ready removal of the compact. Synonymous with Split Die.

Segregation — The undesirable separation of one or more components of a powder.

Semi-Alloyed Powder — An alloyed powder in which the constituents have not reached the completely alloyed state. This is achieved by controlled heat treatment. Synonymous with Partially Alloyed Powder.

Shrinkage — A decrease in dimensions of a compact occurring during sintering (converse of Growth).

Sieve Analysis — Particle size distribution; usually expressed as the weight percentage retained upon each of a series of standard sieves of decreasing size and the percentage passed by the sieve of finest size. Synonymous with Screen Analysis.

Sieve Classification — The separation of powder into particle size ranges by the use of a series of graded sieves.

Sieve Fraction — That portion of a powder sample which passes through a standard sieve of specified number and is retained by some finer sieve of specified number.

Single-Action Pressing — A method by which a powder is pressed in a stationary die between one moving and one fixed punch.

Sintering — The metallurgical bonding of particles in a powder mass or compact resulting from a thermal treatment at a temperature below the melting point of the main constituent.

Sizing — A final pressing of a sintered compact to secure desired size.

Slip Casting — A method of forming metal shapes by pouring a stablized water-suspension of metal powders into the shaped cavity of a fluid absorbing mold, diffusing the liquid into the mold wall, removing the casting from the mold and sintering.

Slip Crack — See Lamination.

Solid-State Sintering — Sintering of a powder or compact without formation of a liquid phase.

Specific Surface — The total surface area of the powder particles per unit mass of powder.

Spherical Powder — Powder consisting of round or globular shaped particles.

Split Die — See Segment Die.

Sponge Iron Powder — Iron powder of porous structure usually produced by the reduction of iron oxide.

Spongy — A porous condition in metal powder particles usually observed in reduced oxides.

Spring Back — See Green Expansion.

Stripper Punch — A punch, which in addition to forming the top or bottom of the die cavity, later moves further into the die to eject the compact.

Subsieve Fraction — Particles all of which will pass through a 45 micrometre (No. 325) standard sieve.

Superfines — The portion of a powder composed of particles which are smaller than a specified size, currently less than 10 micrometres.

Sweating — See Exudation.

Tap Density — The apparent density of a powder obtained when the receptacle is tapped or vibrated during loading under specified conditions.

Transverse Rupture Strength — The stress, calculated from the flexure formula, required to break a specimen as a simple beam supported near the ends and applying the load midway between the fixed center line of the supports.

Upper Punch — The member of the compacting tool set or die that closes the die and forms the top of the part being produced.

Warpage — Distortion which may occur in a compact during sintering.

Withdrawal Process — An operation by which the die descends over a fixed lower punch to facilitate removal of the compact.

—Notes—

RICK DERIN